Sutherland Birdlife

Fraser Symonds & Alan Vittery

with assistance from Dean MacAskill

First edition printed 2018
All rights reserved

Text: © 2018 Fraser Symonds and Alan Vittery
Artwork: © 2018 Fraser Symonds
Photographs: © 2018 Fraser Symonds, © 2018 Dean MacAskill (Pages 12, 92)

No part of this book may be copied, reproduced,
or transmitted in any form or by any means (electronic,
mechanical, photocopy, recording or otherwise)
without the written permission of the authors.

ISBN: 978-1-78808-492-5

Contents

Introduction	v
An Outline of Sutherland	1
Geology	2
Glaciation	4
Soils	4
Climate	5
Sutherland Bird Habitats	7
Coasts	7
Straths	27
Woodlands	33
Farm and Croftland	44
Peatlands	46
Upland	52
Conservation Issues	63
Birds in Sutherland	95
Breeding specialities	95
Migration	96
Key Sites	99
Sutherland Birdlife in Watercolours	107
Annotated Species List	117
Index	191
References	198

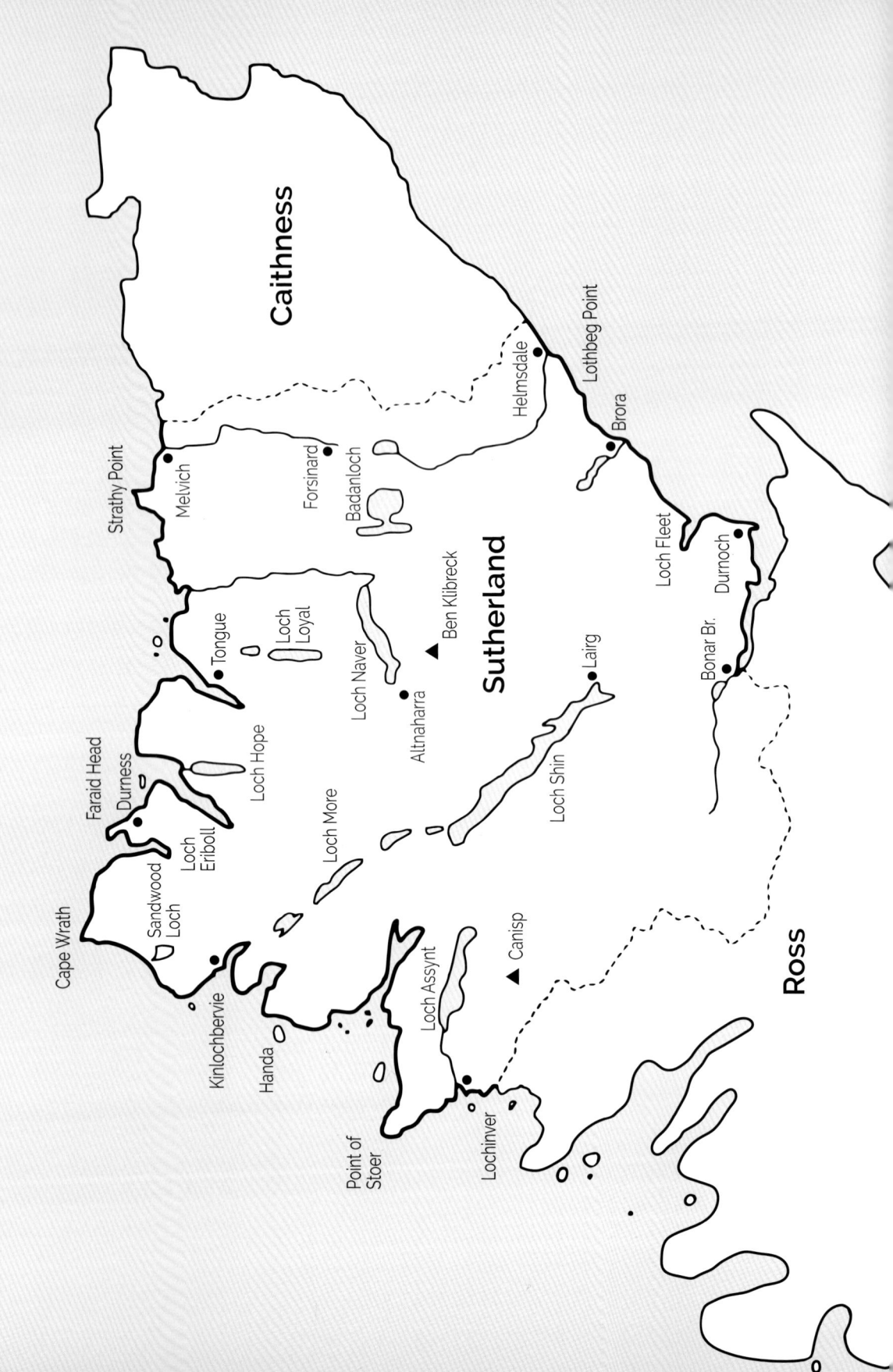

Introduction

The county of Sutherland, in the northern Highlands of Scotland may not boast the most favourable of climates: winters can be cold and long, rain, mist or low cloud is frequent, and when the long summer evenings do arrive, they are often beset by plagues of the infamous Highland biting midge, one of the less attractive forms of wildlife. However, all this is more than compensated for by the beauty of the scenery and the quantity, quality and variety of wildlife which is not readily surpassed anywhere in the British Isles.

Sea cliffs, sandy beaches, small off-shore islands, high mountains, woodlands, lochs, rivers and estuaries, all set against extensive skylines and impressive panoramic views, are here for all to see and rejoice in. The landscape however, is not just a backcloth to please the human eye; it plays an active role in providing the habitats that support a great assemblage of birdlife. Here in Sutherland the habitats exist that provide homes for terns, puffins, red-throated divers, golden eagles, dotterel, ducks, geese and wading birds in their thousands. The list, if not endless, is impressively long. This is a taste of some of the birdlife discussed more fully in the following pages. Some of the species are of course readily seen throughout the year, others are seasonal and some can only be observed with considerable patience or the expenditure of much energy.

The last comprehensive review of Sutherland's birdlife was published twenty years ago in 'The Birds of Sutherland' (Vittery, 1997). Much has changed since then, including aspects of land use and weather patterns, resulting in the decline of some species and the arrival or increase of others. More than 40 new species, mainly rare migrants, have occurred in the County for the first time. Alan Vittery, author of the previous book, was absent from Sutherland for long periods during the last eight years, when the birding baton was ably carried by Dean MacAskill. His extensive fieldwork has greatly assisted the process of determining what changes, some subtle, have occurred during this time.

It is timely then, after twenty years, to reassess the status of all the County's birds, taking advantage of the digital age to present the information in a more attractive format and broaden the scope to discuss how the landscape, habitats and land uses affect the birds which reside in or pass through Sutherland. This book is therefore much more than an update of the 1997 publication. It is a celebration of the richness of Sutherland's avifauna and reflects a recognition of the area's special importance as a remnant of relative wildness in an overcrowded and increasingly developed island. The accompanying illustrations are intended to give visual impact to the text – a text which hopefully guides the reader towards a greater knowledge of what is to be seen and why the particular species of bird occurs exactly where it does. The interplay between birds, habitat and landscape is complex, often changing and often not wholly comprehended, but it is important to attempt to understand, because understanding and knowledge are the prelude to conservation. If the rich, yet vulnerable birdlife of Sutherland is to flourish for generations to come then we need to appreciate how the actions of man both help and damage its prospects. Some conservation issues relevant to the county are discussed in a later section.

A glance at a map of the far north of Scotland begs the question: "Why exclude Caithness?" A history of separate recording is not a satisfactory reason in scientific terms. Curiously though, the contiguous Counties are very different topographically and to some extent, climatically, resulting in surprisingly different bird populations. This is partly due to the flatter interior of Caithness being closer to the sea. During the last Ice Age it was, in fact, a 'nunatak' refuge. Birds have long memories!

Sutherland's more diverse topography, with wild rocky coasts and a mountainous interior in the west contrasting with drier moors and sheltered sandy inlets on the Moray Firth in the east, satisfies the requirements of a wide range of breeders and winter visitors, as well as the human eye.

Vast areas of peat bog render Sutherland one of the least densely populated areas in Europe – further increasing its importance for scarce and sensitive species in our overcrowded islands. Uncountable lochans, rivers, burns and relic woodlands add to the natural diversity, tarnished only by the large-scale planting of alien conifers in the second half of the last century.

The more recent proliferation of wind farms (of very debatable value when sited on peat, which oxidises when fragmented and disturbed) presents a new threat to species like raptors and golden plover, although the disfigurement of many wonderful landscapes is more obvious to the visitor.

Historically, Sutherland figured in the ornithological consciousness mainly as an inaccessible refuge for some of Britain's rarest breeders, like black-throated diver, common scoter, corncrake, greenshank, Temminck's Stint and wood sandpiper. Unfortunately, it was therefore of as much interest to egg-collectors as bird-watchers. Resident ornithologists, such as the late Donnie MacDonald of Dornoch and Dr Ian Pennie of Scourie, were thin on the ground and their activities necessarily localised. Nethersole-Thompson's studies of breeding greenshanks and Derek Ratcliffe's pioneering site work, identifying areas worthy of legal protection as National Nature Reserves or Sites of Special Scientific Interest, made notable contributions to knowledge of the County's birdlife.

Given the proximity of Sutherland to the migrant-rich Northern Isles, it seems obvious that numbers of unusual species must occur in, or pass through the County. Intensive fieldwork over twenty years by Alan Vittery from 1990, including regular 'seawatching' on both the east and north coasts and targeted searches for migrants in favourable weather conditions, was successful enough to put Sutherland on the twitchers' radar. Sadly for them, there is little to tempt strays to remain in any one spot for long, as they might on an island. Many filter south along coasts and river valleys once they have rested and fed.

The spectacle of large-scale migration is more predictable and tends to be concentrated along traditional flight-lines. The NE/SW east coast flyway neatly links the Northern Isles with the Inverness Firth and the Great Glen and is most used in autumn, when many migrants arrive from Scandinavia. The north-west fault, from the Kyle of Sutherland to Loch Laxford, via Loch Shin, is favoured by species heading to and from breeding grounds in Iceland, such as pink-footed geese. Offshore, seabirds can pass in extraordinary numbers, both exiting the Moray Firth (the largest seabird 'trap' in Britain) and through the Pentland Firth in the north.

Despite these riches, relatively few birders visit Sutherland. For those wishing to find their own birds, and enjoy them in spectacular surroundings, there can be few places to match it in the British Isles.

An Outline of Sutherland

Sutherland is one of the largest counties in the British Isles and yet has a population of only a little over 13,000 people. It has landscapes of large scale and is a land of contrasts, comprising the mountainous north west, a vast central plateau of rolling peatland and a narrow band of hills along the east coast, which shelter a low lying coastline.

The main concentrations of people in Sutherland are in the south and east, particularly around Brora, Golspie and Dornoch. There is a road, dual track for most of its length, which follows the coast around Sutherland and provides a link through Caithness. Otherwise, internal through routes within Sutherland are restricted to single track north-south links via Lairg to Laxford Bridge or Tongue and via Kinbrace to Bettyhill or Melvich. This simple outline is enough to indicate that by comparison with much of the rest of Britain this is an area where man and man's artefacts do not dominate and the character of the natural features ensure that birdlife can exist with only limited human interference.

It is not prudent to consider birdlife as a subject entirely on its own, for it is inextricably linked to other aspects of the natural world, most especially geology and climate. It is the type and structure of the underlying rocks which influence the soil type and drainage patterns and these, combined with the climate, determine what can grow and therefore what can live where.

Lewisian gneiss landscape of the north-west, showing Arkle from Ben Stack.

GEOLOGY

There is a decrease in the age of the underlying rock structures from west to east across this region. The ancient Lewisian gneiss landscape of the west and northwest is one of gentle rounded hills, broken by numerous rock outcrops and ridges and countless water filled depressions. It is a predominantly rocky landscape with much unvegetated ground. This is the characteristic 'cnoc and lochan' landscape. Ben Stack, bordering the River Laxford, is one of the highest of the pure Lewisian gneiss mountains, at a height of 721m. Several other peaks, such as Foinaven, are higher but these have a capping of quartzite, which can often lead to fascinating lighting effects.

The three isolated mountains of Suilven, Canisp and Quinag which lie within this Lewisian gneiss landscape are examples of scattered outbreaks of Torridonian sandstone. Moine rocks, created from deposition of sediment in an ancient sea, have been pushed eastwards and now form the bulk of the solid geology which underlies the majority of Sutherland. The landscape here is on a huge scale, with vast expanses of rolling terrain covered by a dense blanket of peat and only periodic rock outcrops. There are, however, many mountains within this zone comprising the same rock type. Indeed the biggest of the Sutherland mountains are Moinian. These include Ben Hope, Arkle, Ben Hee, Ben Klibreck, Ben Armine, Seana Bhraigh and Carn Ban, some of which demonstrate marked erosion surfaces at high altitude. Although the extent of the Moine rock is large, it is not uniform in its cover and there are a number of intrusions of igneous rocks, especially granite. Carn Chuinneag near the Ross-shire border at Ardgay is a notable example. Ben Loyal and nearby Ben Stumanadh are composed of syenite, a close relative of granite.

Mountain landscape of West Sutherland, showing Suilven, Canisp and Quinag, with Foinaven beyond.

Kyle of Tongue and Ben Loyal, one of the few igneous rock mountains in Sutherland.

To the east and throughout much of the neighbouring county of Caithness, old red sandstone predominates. Different age classes of the rock show different characteristics. The oldest are composed of conglomerates, an amalgam of stones resembling coarse concrete. The rounded hills between The Mound and Golspie, or the crag of Carrol Rock above Loch Brora, all on the east coast of Sutherland, provide perfect examples.

GLACIATION

The last Ice Age is an effective starting point for any account of the habitats of the region in that most of the evidence of previous glacial periods was removed by the latest activity. Deglaciation of this part of Scotland was largely complete 13,500 years ago and surface soils and vegetation started to develop. These were not in forms exactly as they occur today, in that subsequent blips in the warming climate led to modifications. Indeed, as 'recently' as 10,400 years ago there were glaciers re-established in some high altitude corries. It was the retreat of this last major ice cap which has formed the landscape as we know it now. The main features are the glacially deepened sea lochs of the West Coast and the widened straths of the east, with numerous mounds of moraine scattered to the sides and at the limit of ice movement. In terms of impact upon vegetation, it is principally the process of subsequent soil formation which is most significant.

SOILS

Most of the soils throughout Sutherland are acidic and composed of much coarse, stony material. The acidity is due to the nature of the underlying rock from which it is derived, together with a climate featuring low temperatures and high rainfall. Both factors lead to poor humus decomposition and an increase in acidity. The distribution of different soil types corresponds with the underlying rock or glacial deposit distribution. The most striking feature however, apparent even to the casual observer, is the dominance of peat, which accounts for up to 30% of the land surface of Sutherland.

Peatland landscape near Altnaharra.

CLIMATE

Weather patterns vary considerably across the county. The mountainous ground in the west and the maritime surroundings both have a major influence. Rainfall amongst the western mountains can be over 3000mm per year, compared with 750 on the East Coast. The wettest period is between August and October, earlier in the east than west. The average daily temperature is 7–8 degrees centigrade and the prevailing wind is south to south- east, but with west to north-west showing a clear secondary peak. During the winter, sea temperatures on the west average 6°C but are about two degrees lower in the east. Temperatures rise only slowly throughout the year and, with frequent onshore breezes in the spring and summer, sea haar is common in the east.

Rugged west coast seascape, north of Kinlochbervie.

Sutherland Bird Habitats

COASTS

There are few places in the world where one can view three different coastlines in one country in the space of one day without leaving dry land. At the far north of Scotland a two hour journey by road is sufficient to move from the shores of the Atlantic Ocean to the coastline of the North Sea, taking in views over the notorious Pentland Firth, offshore islands, sandy bays and deep-set sea lochs en route. This relatively short

East coast landscape looking across Loch Fleet towards Golspie.

journey represents some of the diversity of coastal habitats which can be found along the "three coasts" of Sutherland. In very simple terms the west coast, with its exposure to the frequent winds from the Atlantic, is a rugged, rocky landscape, whilst the east coast, particularly along the east Sutherland coastline is softer and lower, with a greater emphasis on expanses of sand and intertidal estuarine systems. This general difference in landscapes is also reflected in the species of birds which inhabit them.

Cliffs and Offshore Islands
There are numerous islands of considerable variation in size off the west coast of Scotland. With northerly progression they decrease both in number and size, almost to the point of extinction approximately half way along the north coast. Thereafter, with the exception of the Orkney Islands, which are a geographical and ecological entity in their own right, offshore formations are mostly restricted to stacks and fallen rock archways.

None of the islands of the north west are currently inhabited, although most still have derelict buildings which illustrate their past significance as homes for those tending sheep. The grazing flocks long outlasted their flock masters and several islands still support North Country Cheviots or Shetlands to this day. The withdrawal of herds from many of the remainder is relatively recent and the island habitats which exist now owe much to their agricultural background. In addition, the island vegetation is heavily influenced by the underlying acidic rocks and thin peaty soil. It has a natural tendency towards heathland, comprising heather, dwarf shrubs (crowberry in particular) and scrubby willows, with the final mixture determined by the intensity of grazing or the exposure to salt laden winds. The former has led to a more grass-dominated sward where the willow scrub is kept at bay and where wetter ground has a predominance of bog cotton. The most exposed wind clipped heaths balance precariously between decay, through erosion of the fragile surface and survival as a specialised mixture of heathland plants, replicated at few other locations.

Grazing has without doubt shaped the nature of the island vegetation which persists today, but what part does it play in other aspects of island ecology? Not surprisingly, it is a key influence in the fortunes of many other species, nesting birds in particular. In summer, offshore islands can be a haven for seabirds, where protection against land based predators offers safe breeding sites. Some of these birds choose the open habitats away from the surrounding cliffs, where thicker vegetation provides nesting cover. Sheep, goats or ponies can influence the length of the vegetation and proportion of scrub and so determine the presence or distribution of nesting birds. Great black-backed gulls are widespread and fairly cosmopolitan in their choice of nesting site. As long as it is free from predators, tall heath, wet heath, wind clipped heath, coastal grassland and even bare rock will all suffice. The bird's quest for predator free sites appears to outweigh most other considerations and great black-backed gulls even nest successfully on a rocky island off Cape Wrath which is used by NATO forces for testing live air to ground missile systems! The flashes, bangs and flying debris are clearly preferable to uninvited two or four legged visitors.

Arctic and great skuas on the other hand are more particular. These species are not widespread but where they choose to settle, do so in dispersed colonies. Handa Island is perhaps the best known great skua

Wind clipped coastal heath.

colony south of Orkney and Shetland and numbers some 165 pairs. These birds choose the matted mixture of heath and bog cotton to nest in but also show a close affinity to small lochans which can provide essential fresh water for bathing and communal gatherings.

Arctic skuas occur widely offshore but only breed on Handa.

Great skua or "Bonxie".

Storm petrels, nocturnal in their land based activities in order to avoid avian predators, nest in holes, usually underground burrows. These are frequently within heathland or short coastal grassland where access in a hurry is not impeded by long vegetation. This is another example where animal grazing may help with the provision of suitable nesting sites.

It is not just during the summer breeding season that the relationship between mammalian grazers and birds is apparent. Between late October and April each year approximately 800 barnacle geese from Greenland frequent the rough grassland on the islands of the northwest. These avian grazers require short grass to sustain them and were it not for small flocks of sheep which keep it short during the summer this habitat would gradually become rank and cease to be attractive to the geese. Winter flocks of barnacle geese grazing amongst scattered boulders on small islands surrounded by grey winter seas has a sense of correctness, unlike the sight of considerably larger flocks on intensive farmland in the south-west of Scotland.

Barnacle geese from Greenland frequent the fields near Durness and nearby offshore islands in the winter.

It is not so much for their "interior" that these small Atlantic coast islands are well known but for their spectacular cliff lines. Wherever the rock structures are of a type which erode under the influence of water activity to create ledge formations, there are usually colonies of nesting seabirds. The coast of Sutherland is no exception and indeed it is not just the islands which are significant in this respect. Tall, sheer cliffs are frequent along much of the west coast, reaching spectacular proportions at Cape Wrath. The cliff faces at Clo Mor, east of the Cape Wrath lighthouse, contain some of the densest colonies of breeding seabirds. When added to those of Handa island and the scattered colonies which extend eastwards as far as the county boundary with Caithness the combined populations are measured in hundreds of thousands. Clearly, such populations are of international importance and support a high proportion of the total British breeding populations of several species. Numerically, guillemots are the winners across the whole of this region, but a number of other species are in hot pursuit. All the cliff nesters are well established including, kittiwake, fulmar and razorbill. Shags also do well, but especially so where cliff falls have led to accumulations of boulders near to sea level. Razorbills too, favour these areas. Cormorants are more choosy and colonies are both small and well dispersed. Where they do occur, they are either on the flat summits of offshore islands or outer promontories of inaccessible cliff sections.

Breeding seabird colonies on the cliffs of Handa Island. (Photo Dean MacAskill)

Other seabirds not mentioned so far include the species with probably the greatest popular appeal, the puffin and its close relative at the other end of the celebrity profile, the black guillemot. Both occur frequently but, as both are also quite specialised in their habitat requirements, colonies are more scattered than those of other species. Puffins require holes in the ground within easy reach of the sea for their nesting sites. Rabbit holes do nicely but this brings with it a requirement for cliffs topped by suitable turf and inaccessible to ground predators. Small

Nesting kittiwakes with well grown young.

Guillemots breed on exposed rocky ledges.

numbers can be found almost anywhere, but the largest concentrations occur in the west, especially around Cape Wrath, where the broken structure of some of the cliffs allows for better establishment of vegetation and eventually hole bearing turf. A rat eradication programme on Handa island, part funded by the voluntary East Sutherland Bird Group, has allowed puffins to recolonise the main island, having been exiled to the very limited accommodation provided by inaccessible stacks.

Puffin.

Black guillemots are found on coasts with plenty of caves or boulders.

By contrast, the black guillemot is more numerous along the north and east coasts but nowhere do they concentrate in anything more than small groups. They nest amongst boulders or in rock crevices, usually close to sea level.

Away from the main cliff formations, much of the west and north coasts remain rocky. River systems from the mountainous interior break this pattern. River mouths along these coasts are often characterised by narrow, sandy beaches backed by discretely demarcated soft dunes or, alternatively, seaweed-strewn boulders opening into a broad sea-loch. Such lochs, including Laxford and Inchard, lie in spectacular settings, but the wildlife is numerically limited. Herons, scattered curlews and oystercatchers are reliable but the speciality of this area is the otter. The density of these animals in the north is quite high, but finding one is another story. Crabs, flat fish or butterfish are amongst their favourites and, as the ebbing tide reveals new shallows full of various seaweeds, the otters are at their most active.

Firths and Estuaries

One of the greatest contrasts between east and west is evident where freshwater meets seawater. West coast estuaries are little more than widenings of the river mouth at the point of contact with the sea, with a minimal upstream tidal effect. There are few major rivers on the east but of these, two have significant estuarine systems and contribute to the characteristically incised coastline so familiar on any map of Scotland. The Rivers Carron, Shin and Oykel, which all have their origins in mountainous parts of Sutherland, combine forces within the Kyle of Sutherland and then merge with the sea within the Dornoch Firth. It is this water body which demarcates the boundary with Ross-shire to the south. Slightly further north, the River Fleet enters a land-locked basin where ancient sand and shingle bars provide shelter from the open sea.

These two estuaries, especially when combined with their neighbours to the south, exhibit coastal ecological and geomorphological processes of classic proportions and rank along with the seabirds of the north and west as one of the region's internationally important nature conservation assets.

Loch Fleet, from the air..

Dornoch Firth; looking west towards the inner firth.

A dominant feature within the firths and a few neighbouring sites as well, is the presence of extensive intertidal flats. The tidal range on this coast is four metres at its maximum and combined with the shelter provided by the deep indentations, produces great expanses of tidal flats which range in structure from the sloppy mud of the upper reaches to firm sandy mud nearer to the scouring effects of the main channels or open sea. This variety of formation, together with the richness of incoming waters and shallow, productive seawater, leads to an abundance of life forms within the mud which is better seen than described. Suffice to say that the densities of shellfish, marine worms, crustacea and amphipods living within the top few centimetres of the mud would, in human terms, make our densest cities look deserted by comparison.

Not only do invertebrate fauna thrive here but marine plants also do well. These are specialists, capable of withstanding total submersion in salt water followed by exposure to desiccating sunlight. Not surprisingly the diversity of species of such tolerance is low but where they occur, they often do so in abundance. The principal species on these flats are species of eel grass or Zostera. These thread like marine grasses grow during the summer months to create patchy but dense carpets upon the

Intertidal mudflats such as those at Dornoch provide valuable feeding grounds for thousands of wintering shorebirds.

firm sandy mud sediments. These plants play an important role in the ecology of the estuarine systems for they act as the principal food supply for migrating wildfowl in the autumn, as they arrive from their summer breeding grounds. The summer growth is there for the cropping by grazing wildfowl, otherwise it is lost to the autumn storms which simply rip it from the mud and deposit it in lines along the storm tide line.

Wigeon is the main species to benefit. The peak arrival of birds from Iceland occurs in mid October and, in years of good Zostera production, numbers stack up to reach levels in excess of 35,000 within the Moray Firth. The Dornoch Firth alone may support 17,000 for a short time. The food stocks will not sustain this number for long and in any case the Zostera is in only temporary supply in the these northern latitudes. The majority therefore move through, en route to southern Scotland

Wigeon reach peak numbers in the Dornoch Firth in October and November.

or further south still. By mid December wigeon flocks have declined to levels which can be sustained through the cold winter period.

This pattern of autumn passage is not the norm for estuarine birds on the Scottish east coast. Indeed for the majority of species it is the opposite. Peak numbers of waders occur in late December or January. At first sight this seems illogical, for it is at this time that conditions on the shore can be at their most severe. There is method in this apparent foolhardiness however. Consider the cases of dunlin, knot and bar-tailed godwit. All those that occur in Britain during the winter breed in arctic latitudes; dunlin and godwits from arctic Scandinavia and Russia, knot from Greenland. In late July when the young of the year have just grown new feathers, they set off for their wintering quarters in Scotland. They arrive without further ado at estuaries throughout the country at the time of year when their small invertebrate prey in the mud has just bred and densities are at their annual peak and most readily available to the inexperienced young wader. There are few adult birds around and competition is not great. They therefore have the chance to build up their reserves for the winter ahead. Meanwhile the adult birds have just completed the arduous task of raising broods. Their flight feathers are twelve months old and in poor condition and they have to moult out of their summer body plumage. This is very demanding on energy and combined with this they need to vacate the arctic before it freezes. Unlike their offspring they take their time over the journey and instead of crossing the North Sea in one go, journey down the Scandinavian coastline stopping regularly at the many intertidal sites available until they reach the German or Dutch sections of the great Waddensee. The tidal ranges here are low and the sand flats enormous in scale. This strategy enables the adult birds to feed voraciously on these productive mudflats whilst they are moulting their feathers and starting to rebuild their reserves. They cannot stay on these Continental coasts in vast numbers during the main part of the winter because the continental influence causes them to freeze more readily than the island shores of Britain. Come early winter the replenished adults cross the Channel to the east coast and disperse throughout the estuaries of the UK. Many are known to move steadily northward as the winter progresses. The reason for this remains unclear, but perhaps to be nearer to the breeding grounds come departure time in spring. Thus the pattern is set and the number of wading birds on the mudflats of the north Highland Firths gradually rises as the winter progresses and the proportion of adult birds rises with them.

Juvenile knots.

Large flocks of waders are a common sight on the estuaries of the south-east in winter.

There are of course many waders present throughout the year. There is a strong population of oystercatchers which breed both on the shores of the Moray Firth and inland. These are supplemented by those which breed in Shetland, the Faroe Islands and Iceland. There is a similar pattern for curlew and possibly ringed plover.

Oystercatcher flock.

Ringed plovers breed throughout the county but winter mostly on the east coast.

Dunlin from Scandinavia winter on the shores of the Moray Firth.

Curlews alighting in Loch Fleet.

The key element in this complex avian world however is the presence of the mudflats which contain a rich food supply and remain relatively ice free during the winter, together with a selection of safe and undisturbed sites on which to roost and save energy whilst the flats are covered by the high tides.

The wading birds are just one piece of the jigsaw. There are many species of wildfowl which breed in Iceland and other arctic areas and the North Highlands acts as winter host to many of them. Greylag geese start to arrive in October and on clear autumn days it is quite easy to spot skeins several miles out over the sea before they reach land. Few stop immediately, preferring to head to their favoured stopover points in Caithness or near the shores of the Moray Firth. Agricultural land provides the best feeding but the mudflat habitats are ideal roosts, especially within sheltered bays. Whooper swans follow a similar pattern but numbers for this species are measured in tens rather than the thousands of the previous species. In some years however, grain harvests are late and much grain can spill from the heads prior to combining. Occasionally this can occur within some fields on a scale which leads to a bottle-neck of both swans and geese, as new birds arrive faster than the early arrivals move on.

The greylag geese which winter in Sutherland are very closely correlated with the sheltered intertidal bays of Loch Fleet and parts of the Dornoch

Firth, where they roost at night, but it is probably the adjoining agricultural land which has a greater relevance to their distribution. There are two discrete populations here although, morphologically, they are indistinguishable. Native bred greylags frequent Loch Fleet and the fields to the north near Golspie. These birds are present from August to February and maintain a constant flock of about 500 birds. A second group of similar size gathers in the inner Dornoch Firth in August and September but disperse soon after. Numbers rise rapidly in October as birds from Iceland start to arrive. These also gather at Loch Fleet and can boost numbers to 5000 on occasions. They also favour the sheltered bays of the Dornoch Firth for roosting and feed regularly on nearby fields. Three thousand or so birds are not uncommon here but these individuals mix little with the native birds.

Greylag geese are resident in the county but peak numbers occur in late autumn.

The sheltered and shallow sea of the Moray Firth off the East Sutherland coast has traditionally been favoured as a wintering area by migratory sea duck but here lies a topical conservation question to which there is no current answer. During the 1970s and 1980s the waters between the Dornoch Firth and Loch Fleet were frequented by rafts of eider ducks, long tailed ducks, and common scoters, with occasional rarities such as king eider or the North American surf scoter amongst them. Flocks would drift offshore as conditions permitted and so counts varied considerably but at peak times flocks of 10,000 long tailed ducks and several thousand eiders were not uncommon. By the early 2000s numbers had crashed and sightings of all but small parties of these species were noteworthy. Today there is an indication numbers may be recovering, especially for eider. Despite lower numbers, visits to this part of the coast in winter are still rewarding, especially if the sea is calm and visibility good. All three of our wintering species of diver occur here regularly and on occasions are often accompanied by parties of slavonian grebes.

Flocks of eiders can be seen along the east coast.

Long-tailed ducks gather in small flocks in the Moray Firth during winter.

Dunes and slacks

Sand dune systems are characterised both by their soft mobile sand and also two dominant grass species. Anyone who has walked through a sand dune will be familiar with the strong stabbing points of the marram grass and, near the strand line, the loose seed heads of the tall and elegant lyme grass. The former is the principal agent in binding the mobile sand into a more static form due to its complex and voluminous network of fine roots. Once "fixed", dunes soon become colonised by other species, but only of a limited range. Mosses soon carpet the surface and sand sedge may run in distinctive straight lines as new shoots emerge from nodes on the underground runners. Rosebay willowherb may also invade in bulk, but species which indicate a less disturbed history include the bright yellow kidney vetch. It is not just the plants which give these dunes their character. Yellow and black striped snails are numerous and wrens, stonechats and reed buntings often perch on the stems of tall lyme grass. In winter the seed resource within the dunes may allow flocks of other buntings to gather. Migratory snow buntings are often characteristic, but in harsh weather, yellowhammers too.

Immediately inland from the dunes, there is frequently a low lying zone, often with seasonally lying water. This quite commonly retains some

Yellowhammers often move to the coast in winter.

Windblown sand trapped by marram grass to form embryonic dunes.

The slacks on the landward side of dunes often flood in winter but are rich in plants and birds in summer.

Temporary tidal pools flood on spring tides and attract wildfowl to feed.

elements indicating its brackish past such as Baltic rush or saltmarsh rush but the principal flowering species may well be dependent upon the freshwater influence.

Such habitats are a haven for botanists, but there is much to interest the ornithologist too. The damp areas are frequented by breeding waders, particularly redshank, curlew, oystercatcher and lapwing. Shelduck which breed in dune habitats also enjoy the fresh water that these dune slacks can provide. As the numbers and distribution of nesting lapwings continue to shrink throughout much of Britain, for reasons as yet not fully explained, this habitat is becoming increasingly important.

STRATHS

Rivers and floodplains
A glance at a map of Sutherland will show both a multitude of water courses and a curious imbalance in the water flow pattern from east to west. This is due to the greater abundance of mountains in the west, which create a watershed rather close to the west coast. Consequently, along the Atlantic coast, the result is a number of short, fast flowing, narrow rivers derived from relatively small catchments. Few of these rivers exceed 10 kilometres in length. By contrast there are far fewer rivers flowing to the North Sea, but each of these has a vast catchment and a fine cobweb of branches and minor tributaries in the interior. As the rivers progress

Upper reaches of the River Cassley.

nearer to the sea they merge to form major aquatic arteries. One of the principal systems in the south east of the area is the Kyle of Sutherland which meets the sea at the upper reaches of the Dornoch Firth at Bonar Bridge. This Kyle comprises the waters of the Oykel, Cassley, Shin and Carron, all of which have significant tributaries themselves and all have origins in the mountains of the Moine zone of east Sutherland. Moving northwards from the Dornoch Firth one encounters the wide straths of the River Fleet south of Golspie, the River Brora and the Helmsdale. Along the north coast the principal waters are the Halladale and Naver, although there are many smaller systems in addition. Away from the west coast and the rivers supplying the Kyle of Sutherland, the majority of the rivers have their origins in the peatlands of the centre.

All the rivers of the area are subject to frequent spates, their flow rates responding quickly to changes in rainfall. After heavy rain the water levels rise rapidly, but this increased flow is not maintained for long and soon subsides. Decreasing quantities of lying winter snow over recent years has tended to lead to lower summer flows than in the past.

For much of their length these highland rivers flow through open country. They have as their base either solid rock, especially in the upper reaches, loose gravels or softer peaty sediment. Most of this open country is grazed, either by sheep or red deer. Consequently there are few areas with wooded or tree lined banks and many areas have soft, eroding banks. The major rivers follow glacially enlarged valleys and in some cases, especially the Kyle of Sutherland and the Helmsdale, have wide, level flood plains. In the former, much of the ground is ill-drained and is a habitat rich in marsh or swamp plants. The diversity is maintained by seasonal grazing which helps to prevent the build up of rank, tussocky vegetation. The wetness of the ground also limits the intensity of the grazing and patches of willows and alders thrive.

The key features of the wide flood plain of the Kyle are the frequent patches of riverside sedge beds and marshy areas dominated by common reed, marsh cinquefoil and bogbean. The mixture of open water, dense reed beds and scrubby patches is attractive to wetland birds such as herons, teal, wigeon and in winter, groups of whooper swans. Waders use the river channels as corridors between the coast and their inland breeding sites. Oystercatchers, greenshank, redshank and lapwing are all found here and some stay to breed on the riverbanks. Snipe too, do well in the marshy ground.

Kyle of Sutherland marshes at Culrain.

These major rivers form the main drainage arteries but, numerically, it is the upland burns which dominate the flowing aquatic environment. These vary considerably in character, often as a result of the location of their source. Burns and rivers originating in rocky, high ground will be steep and fast flowing. They are likely to have hard rocky beds and banks and may flow through steep sided narrow gorges. Those with upper reaches deep in the peatlands of the interior will have an appearance not unlike an aimlessly wandering canal. Here the flowing water has cut a channel with an even profile through the soft, peaty soils. As the ground is generally of a gentle gradient, such rivers are often relatively broad and meandering.

The meandering channel of the upper reaches of the River Brora.

The banksides of these two forms of watercourse differ, as does the wildlife which frequents them. The fast flowing burns of the uplands carry little silt and the water is often clear. These are ideal conditions for larval stages of insects, which in turn form tasty meals for trout or young salmon. They also are the staple diet of riverside birds such as the dipper or grey wagtail. There are few Highland burns which do not have resident pairs of the brown and white 'semi submersible' birds scattered at regular intervals along their length. The sight of a dipper jumping off a rock and disappearing below the surface into the fast current, only to appear moments later on another boulder in mid stream is one which can occupy many a pleasant hour. More usual however, is the sound of their short metallic call as these energetic little birds fly at speed up or down the course of the river.

It is quite common for these hill burns to flow through narrow ravines or gorges. Often short, and often in otherwise open terrain, such places hold hidden gems. The steepness and rocky ground may encourage hungry sheep or deer to look elsewhere for sustenance and so the vegetation is richer and more luxuriant than more accessible ground. It is also frequently damp and not well lit by direct sun. This is the home of moss covered rocks with saxifrages, ferns, wood sorrel and a surprising array of flowering plants. Sometimes scattered birch, rowan or aspen

Mixed native woodland bordering an upland burn.

trees indicate a former era when woodland was more widespread than today. These more wooded sites can be a secluded wonderland of subtle colours created by mosses and lichens. It is not necessary to know the names of the multitude of these simple plants in order to admire the delicate complexity of their shapes and structures. It is all too easy to be in a hurry and walk past with injurious footsteps but a little time spent absorbing the detail of these secluded watercourses is time well spent.

By contrast, the water within the slow flowing peaty burns is a murky brown as it carries with it much fine peat stained material. This water is more acid than its clear counterpart of the mountains and has a different ambience. It is often the openness and the bankside vegetation which provide the character. The openness itself provides the key clue to the type of plant life one might encounter. This is sheep country and anything palatable has a short life expectancy. Short cropped grass with an abundance of rushes fringing the waterway is typical and widespread. Areas with impeded drainage are well worth closer inspection, for sheep favour these less. The taller growth might include marsh thistles, or the colourful heads of the yellow flag iris. At ground level the diversity of the flowers will depend on the amount of nutrients being carried to the surface by water seepage. In impoverished situations, the multiple heads of cotton grass and scattered purple flowers of lousewort may be the main components but where the ground yields up its minerals more readily there should be a mixture of sedges, dwarf rushes (such as the black bog rush) and perhaps some insect eating butterworts, with their sticky fleshy leaves and deep blue flowers.

Rivers in these habitats are more suited to wildfowl than the faster flowing versions. Mallard can of course be encountered more or less anywhere where it has rained in the last twelve months. Teal are more particular, but not a lot. They enjoy the back waters of the slower flowing rivers especially where there is ample vegetation to hide in. Wigeon are often thought of as a bird of the coastal mudflats in winter but these are the visitors from Iceland. There is also a small native population and in the summer they frequent the peaty rivers and burns of central Sutherland, where they breed amongst the rushes. Red-breasted mergansers are similar in that they too can easily be seen in autumn and winter in small parties, either on the sea or on sheltered coastal inlets, but in summer move to the river systems to breed.

Another example of a migrant bird with a small resident population that frequents the straths and waterways during the breeding season is the

Female wigeon with ducklings.

greylag goose. The large flocks of winter comprise birds from Iceland and, depending upon the location, some of the native birds too. The latter generally winter in north west Caithness or the fields near Golspie, on the east Sutherland coast. During February they start to disperse from their main winter quarters and move in small groups up the main river systems, the Helmsdale in particular. They gather in parties on the riverbanks, building up their body condition by grazing on the sprouting cotton grass and rushes. Some will remain to breed and rear their goslings on the rivers. Others continue into the interior and nest amongst the lochans of the peatlands. The river nesting birds will remain here until their young can fly and then, without delay, head coastwards and be back on station at Golspie before the first frosts.

WOODLANDS

Native woodlands (birch and pine)
'Native woodlands' is a much used term and with a variety of meanings. The strictly literal meaning is a woodland of entirely natural origin with no subsequent modification by man or man's accomplices (sheep). By implication only trees of native species will occur here. Just how many of these woodlands exist is a difficult question, because even if the wood appears to be of natural origin, who is to say that it has not been manipulated at some stage in the past. A more practical definition is a woodland comprising only woodland plants which are native to the locality and which has the general appearance and structure of a natural wood. Evidence of some past planting, coppicing or pollarding matters little if the basic elements are intact. The presence of introduced species, oft termed exotics or aliens, certainly reduces the claim of any wood to be native. A predominance of the latter renders a claim totally void and alternative labels come into being. 'Deciduous' or 'broadleaved woodland' best describes woods comprising hardwoods of recent origin, or in the case of conifers, 'plantations'. In such cases the natal origin of the tree components is not clear and the terms apply equally to plantings of introduced species as to those of natives.

Does this biological correctness really matter? In a word, yes. To justify this answer might take a little longer and will venture beyond the scope of this book. However, Charles Darwin hit the nail on the head with his investigations into natural selection. At the species level, the prolonged process of natural selection gradually adapts the individual members of that species to fit into their surroundings. A habitat is simply a collection of species all living together. If all members of that community are native to the locality and have evolved together, the whole entity is well adapted and thrives. The chances are it will have a great diversity of members and therefore a rich and stable environment. A native woodland is no exception. It will comprise a variety of tree species, together with a large assemblage of plants, mammals, birds, insects etc, all of which depend upon each other to some degree.

This stability may not necessarily suit man. Native tree species may grow slowly, they may be ill-shaped, they may have wide growth patterns which means only few will grow per unit area. An introduced species may meet man's needs better. It may grow faster, straighter, closer together, have fewer unnecessary branches; whatever. This is not to say this introduced species is not suited to its new environment. It may be, but at what cost? It will be a newcomer, it will not have had time to develop a full association with other members of the community and reach harmony. It may grow faster than its native companions and shade them into submission, it may produce seedlings faster and gradually take over, it may bring unknown diseases or it may fall victim itself to diseases the native species have become accustomed to. All in all it leads to a reduced diversity within the habitat and a lack of stability. This matters little where the objective is purely to utilise particular characteristics of the introduced species and indeed should be exploited to advantage. However, everything has its place and the place for introduced species is not where they will reduce the diversity and wild richness of near native woodlands. How true is the expression 'variety is the spice of life'.

Today, native woods within Sutherland, or more correctly, woodlands of native species and character, fall into four main categories. These are native pinewoods, of which there are few and all within south east Sutherland, oakwoods, birchwoods and wet alder woods with willow. All these represent a relatively small proportion of former woodland cover in the area but causes and timing of the decline are not clear. Clearance by man for agriculture and firewood, combined with subsequent degradation through deer and sheep grazing are undoubtedly significant factors but so too is climate. What is clear from old maps and historical records is that what we refer to today as native woodland has a long pedigree and is known to have been existing as mature woodland in the mid eighteenth century.

Wet willow and alder woodland at The Mound.

The current versions vary in character depending upon their location and degree of exposure. Many are small remnants of scattered ancient specimens on inaccessible rocky slopes or in ravines. Others thrive in sheltered river valleys and yet more do well on open hillsides along the main Straths.

Mixed species native woodland.

Birch woodlands

Birch is an adaptable, pioneer species capable of seeding freely and establishing itself with relative ease in places ranging from damp heathland to boulder strewn hillside, including most places in between. This gives it a competitive edge in areas where soil and climatic factors conspire against slower growing, more robust species which require an element of shelter and rich mineral based soils. Thus, birch is the dominant tree in Sutherland. Rowan usually occurs alongside birch and is also a good pioneer species, colonising new ground or re-establishing on areas now devoid of trees with relative ease. It does not though, provide the density of canopy that birch can achieve and is commonly numerically subservient. Many straths contain areas of birch wood on the hillsides or along the tributary burns flowing into the main river. If there are sheltered areas with long established woodland cover which has led to the development of rich soils, other species may also be included. Hazel, gean (wild cherry), oak and occasionally ash are the principal additional species and if the conditions are favourable, may reach dominance. These species are not typically pioneers and require other species to do the preparatory work. Once established though, some of these are long lived trees and will outgrow birch over 80 years or so. Oak dominated woodlands are not common but do occur on the sheltered shores of the Dornoch firth and in patches along several of the straths in the east of Sutherland. The specimens here include both the pedunculate oak and sessile oak, although superficially there is little to distinguish the two. Oak woodlands also occur on the west coast where the climate is a little more equable. To compensate for the increased mildness there is often greater exposure to wind and rain and the western oak woods comprise trees of a characteristically stunted and twisted form. This does not seem to discourage them from reaching great age however.

Pure birch woodlands are a phenomenon of the far north because, although birch is a very widespread tree, it rarely grows for long as a dominant species and is usually accompanied by many other species in the canopy. It is usually these that produce the overall character of the woodland and, surprisingly, because of this there is no botanical

Mature oak woodland.

Native broadleaved woodland on slopes below Quinag.

classification for a birchwood as such. In Sutherland birch readily reaches maturity with no other tree species for company. Such woodland often has an open structure and comprises trees almost exclusively of downy birch.

Broadleaved woodlands are rarely quiet. Even in mid winter groups of coal tits, often accompanied by blue tits, great tits, long-tailed tits and chaffinches chatter away quietly as they pass through the canopy in search of sheltering insects. Bullfinches, redpolls, siskins, treecreepers and great spotted woodpeckers are all resident and can be seen or heard at any time in most types of woodland. Woodcock are also numerous in winter, especially in patches of birch woodland with an open structure where the birds are free to move with ease between the trees in search of earthworms and other ground living animals.

By spring, the diversity of birdlife increases dramatically. Willow warblers are perhaps the commonest and most widespread songbird and there are few areas of wooded cover which do not issue forth the cheerful, if perhaps rather repetitive, song. Other warblers are much more scattered. Wood warblers favour mixed woods which have amongst them some old, large specimens, particularly oak. Blackcaps are rather few and far between and whitethroats a bird of scrubby habitats. Redstarts arrive during May and can be seen in a variety of woodlands. They like open habitats where there is plenty of light reaching the woodland floor, but their primary requirement for breeding sites is a supply of old, decaying trees with holes to nest in. Redwings breed erratically in patches of scrubby woodland within Sutherland and their characteristic spring song is often the only way in which their presence is betrayed.

Native pinewoods

Pinewoods come in all shapes and sizes but some stand out from the crowd as woodland sites par excellence. These are the pinewoods which are native in every sense of the word and they are few in number. Pine trees have an ability to adapt to their local surroundings to the extent that they become genetically separable. Thus the geographical origins of pine trees can be identified with some degree of accuracy by genetic analysis. From this it is clear that there remains within the Highlands of Scotland a small number of pinewoods which have been in existence since the last ice

Great spotted woodpecker; a common bird of native woods.

Native pinewood, with birch amongst the pines.

age and are a true part of the ancient wood of Caledon. It is true that some, or all, may have been exposed to some form of human attention but the essence of their nativeness is intact. This is immediately apparent from the size and shape of the gnarled old tree stock, the open structure of the canopy and the luxuriant growth of the characteristic ground flora. There are three such woods within Sutherland and all are at the natural northern limit of pines in the current climatic period.

The Sutherland examples differ from their relatives in Ross-shire or Speyside, largely in the detail of the ground flora, but the generalities are the same. The woodland floor has an assemblage of common species which form a dense cover. The three dominants are heather, blaeberry and cowberry, with crowberry additional in places. These are accompanied by a rich mixture of feather mosses, few of which have common names. Hylocomium splendens, Pleurozium schreberii, Hypnum cupressiforme, and Pseudoscleropodium purum are the

A rich moss and dwarf-shrub ground flora within a native pinewood.

constants and the delicate and much rarer Ptilium crista-castrensis acts as an indicator of the long established and undisturbed nature of the woodland cover. Within this basic ground layer there will be specialities. Amongst the orchids, there are two. Lesser twayblade is the commonest, usually hiding in the shade of heather and creeping lady's tresses are widespread in the east but very local in the wetter west. Also the tiny and aptly named twinflower may form mats of delicate pink bells during June.

At greater heights above the ground, holly and juniper may form a low growing sub-canopy and birch and rowan may also be present. This then, is the appearance, but this is just the start, as native pinewoods are home to some of the species which have a traditionally Scottish association. Crested tits, crossbills, (common and Scottish), capercaillies, red squirrels, pine martens and wild cats are all part of the native fauna, although they may also occur outwith the native woodlands. Most of these are timid beasts and do not readily present themselves to the casual observer. Crested tits and crossbill are the easiest to find as they usually give themselves away with their calls. Crossbills are noisy and are easy to detect throughout much of the year, but because of their passion for the tops of pine trees are not always easy to locate. Crested tits in Sutherland are at the northern limit of their range in the UK and are not numerous. Current native woodlland expansion projects may result in a small range expansion. (See below)

Capercaillies were relatively widespread in the native and plantation pinewoods of south east Sutherland up to about 1990. In common with the rest of their range, they have undergone a serious decline and are now thought to be extinct in the county.

The decline of the capercaillie is a sad tale but more happily the future of their habitat looks secure. Over the same time period as the decline has been in progress, so too have efforts to reverse the reduction in area of pine woodland. Considerable effort has been expended by landowners, forestry organisations and conservation bodies to extend native woodland sites through natural regeneration from the existing trees. This has usually been by means of exclusion of deer, to enable seed

Lichens and moss hummocks forming the ground flora within an old pinewood.

Long-established plantation pinewood at Loch Fleet National Nature Reserve.

fall to lead to a new generation of seedlings which are free from browsing pressure. At the same time, new 'woodlands' have been created on suitable ground by planting seedlings from native seed in a manner which replicates the natural pattern. This bodes well for the future in that artificially created sites which have the characteristics of the natural product can also be of high nature conservation interest.

There are examples of plantations of Scots pine which have been in existence on favourable ground for sufficiently long for the fauna and ground flora to have acquired some of the characteristics of native pinewoods. The regular spacing and uniform growth pattern of the pines are the only obvious clue to their planted origins.

Commercial plantations

The low lying ground twixt improved grazings and the hill is often of marginal agricultural value. This is often the favoured ground for forestry expansion. It frequently replaces rough grassland, sometimes with elements of heath amongst it. This is not especially productive ground for wildlife either and is generally an appropriate location for forestry development. Commercial plantations usually have an emphasis upon Sitka spruce as the primary timber crop but not exclusively so. Lodgepole pine is frequently used as a nurse crop in exposed areas and Scots pine can also be used as the final product. Forests planted on prepared ground at densities of over 2200 per hectare, or 1 tree every

Male black grouse at the lek.

metre, lead to very dense canopy cover during the development stages and a resultant decrease in wildlife diversity within. However, no wood is completely devoid of interest and open areas usually support roe deer, and a variety of woodland edge plants. Small birds such as coal tits and chaffinches utilise such sites. During the early stages of development, within the first ten to fifteen years after planting, the grassland and heather vegetation between the young trees thickens and provides a habitat, albeit a temporary one, for a number of significant species of birds. In some cases this is in response to an increase in abundance of small voles which attract birds of prey such as short-eared owls and hen harriers. The thick ground vegetation also provides good nesting cover. Black grouse is another species which responds positively to such changes. This is a bird of woodland edge where it can benefit from the cover of the trees but also come into the open for feeding and displaying. In common with its relatives, the red grouse and capercaillie, it is having difficult times at present, due in part to cold and wet springs which are causing high chick mortality, leading to a significant population decline. Black grouse were widespread in Sutherland in times past, but are sadly quite uncommon today. The scattered strongholds in young plantations are therefore welcome news, especially as there is growing awareness as to their requirements and some plantations are actively designed or restructured to their advantage.

FARM AND CROFTLAND

There is little productive agricultural land within Sutherland. What there is lies on the low lying ground along the east coast from Brora southwards to the Dornoch Firth. This zone overlaps with the area of underlying Old Red Sandstone rock with soils comprising productive humus-iron podzols or brown forest soils. They are generally light, with a high sand content and are free draining. Arable cropping, particularly malting barley, together with potatoes and beef rearing takes a significant proportion of the better quality land. Sheep though, are widespread throughout.

Crofting is more thinly scattered and reflects the distribution of alluvial soils, patches of peaty podzols or brown forest earths along the main straths or around some of the coastal fringes. The principal production in these areas is stock, with an emphasis on North Country Cheviot sheep and associated fodder crops. The size of units is much smaller and production levels lower. Consequently there is still an emphasis on traditional methods of extensive production and this is reflected by the differences in the nature of wildlife which can be found in crofting and the more intensively farmed areas.

West coast crofting township.

The gradual shift from hay making to silage which has taken place over the past few decades has simplified animal husbandry but has few environmental advantages. The richer grassland has little botanical diversity and is unsuitable as breeding habitat for waders or small birds because of the thickness of the vegetation and the early harvesting dates. It does, however, follow that better quality grass feed is available for herbivorous birds in the winter and migratory geese have not been slow to adapt to this. The distribution of wintering goose flocks closely echoes that of silage and spring corn production. Icelandic greylags are the main species but pinkfeet occur in large numbers on spring passage and are starting to overwinter in increasing numbers.

Crofting land is a smaller scale version of the farmland, where field boundaries are generally smaller and production methods less intensive. The ground is usually less forgiving and cropping has to fit around the landforms and natural obstructions rather than fight against them. The wildlife value is often higher as a result. Many crofting townships include areas of open hill, usually as part of the common grazings. This land may well be heavily colonised with gorse scrub. Indeed, in parts of the east coast in particular the hillsides may become yellow during May and into June as the whins come into flower. Gorse itself, although attractive in colour has little inherent wildlife value. It is little grazed, especially when well established, and so its value lies largely in its thickness of cover and the nesting opportunities it offers to small songbirds. Greenfinches, song thrushes, linnets and stonechats are all birds associated with this habitat, but only where suitable feeding is available as well. One other characteristic of gorse covered hillsides, whether this be good or bad, is the preponderance of rabbits. The gorse provides cover, but also gradually alters the soil characteristics to make it soft and dry and thus provides ideal ground for excavating burrows. Where grass grazing is also available rabbits reach numbers of epidemic proportions. This is often bad news to the crofter or farmer but in addition, if this coincides with grassland or woodland habitats of conservation interest significant damage can be caused. On the plus side however, buzzards are not short of food and the density of breeding pairs throughout the low lying parts of Sutherland are amongst some of the highest in the UK. In March and April the mewing cry of displaying birds is a distinctive sound of the area

Crofting township of Elphin, showing enclosed in-bye ground and hill grazings.

which provides a great deal of the areas character. Later in the summer this is replaced by the plaintive cry of newly fledged birds which pester their parents for food incessantly. The long-suffering adults appear to tolerate this behaviour throughout the winter and only take steps to terminate it as the next breeding season starts to approach. Red Kites are now starting to become established in Sutherland as they gradually extend their range north from their Inverness and Ross-shire strongholds. These birds too do well in areas with abundant rabbits, although mostly scavenging on remains of animals which have succumbed to other causes of death.

PEATLANDS
The Moine rock which underlies much of central Sutherland and a good part of western Caithness is acid in nature. In combination with an oceanic, or as we know it, a cool and wet climate, this leads to an exaggeration of the effect of the acidity, to the point where it reduces the breakdown of organic material leading to an accumulation of peat. The landscape of this zone is one of open, gently rolling terrain, with vegetation dominated by the effect of a blanket of peat. It is this habitat which gives the counties of Caithness and Sutherland their special character and outstanding wildlife value. This is a habitat on a grand scale, covering some 401,000 hectares. To put it another way, it occupies a stretch of land approximately eighty kilometres by thirty five. This amounts to a notable proportion of the world's resource of blanket bog.

Other sizeable examples occur along the Alaskan Panhandle, the Northeast Asian peninsulas of Kamchatka and Hokkaido, a small part of the Canadian eastern seaboard and the southern tip of South America.

It is not just a question of extent but also one of quality. This is an ancient habitat, fundamentally unchanged for more than 5000 years and it is still alive and developing. Many blanket bogs throughout the world are dead. They have lost the essential surface layer of bog mosses which are the powerhouse of the peat formation process and thus they have ceased to grow. Not so in the north of Scotland. Bog mosses of the genus Sphagnum are widespread and prolific and continue to grow, die and decompose in a continuous cycle, to the extent that the bog thickness is growing at a rate of one to two millimetres per year. For this to continue, two key requirements need to be met; a plentiful and regular supply of surface moisture and a lack of physical disruption of the bog surface. The former is not much of a problem for this exposed land and as long as the intrinsic value of this habitat continues to be recognised, nor should be the latter, although the siting of windfarms on peat has not helped.

Many areas of blanket bog contain fragments of tree roots. Analysis has shown these to be pine of approximately five to seven thousand years of age, which have been preserved within the peat. They are an indication of a previous climatic age when it was cooler and drier and pine was widespread. A change to warmer and wetter weather at about this time caused the demise of pine forests in these open, central regions and provides a reasonable estimate of the age of the current bogs which have changed little since their formation. Some pine stumps can now be found buried under several metres of thick peat. The understanding of the history of these peatlands is really only the icing on the cake for it is the cake itself which is of greatest interest.

For anyone with any interest at all in natural history, the peatland habitat, both the whole landscape and the close-up detail is fascinating. The true blanket bogs ocurr on extensive, level ground due to the physics of water movement. From ground level it is often difficult to appreciate the complexity of their structure but by gaining a little altitude it all becomes clear. The bog surface is an intricate pattern of pools or dubh lochans of irregular shape. The proportion of firm ground is often relatively low and each bog has its own distinctive pool geometry according to the precise nature of water flow patterns.

At close range the predominance of water saturated mosses will be evident, but growing through them will be a mixture of cotton grass, cross-leaved heath and deer grass. Stunted heather is widespread but only reaches dominance on drier slopes. In mid summer the bog reveals its true diversity. The surface of the pools become crowded with the white, star-shaped flowers of bogbean, the edges with spikes of bright yellow bog asphodel and the more open peat, with diminutive, carnivorous sundews.

The moss flora is not confined to semi floating carpets of sphagnum. Some species grow into sizeable hummocks and one in particular, the grey coloured woolly fringe moss, does so with such alacrity that the landscape becomes speckled with mossy humps. Unusually for mosses these are dry to the touch. The oldest examples often support crowberry which enjoys a respite from the surrounding wetness. These humps also provide lookout points for small birds as well as convenient feeding stations for larger ones.

There are also trees here, about ankle high when growing strongly. These are the dwarf birch, more commonly associated with arctic tundra. They are perfect miniature replicas of full size birch trees, save for a rounder leaf shape.

The semi aquatic habitat forms a rich breeding ground for insects, which in turn attracts insect eaters. These include newts, lizards and frogs as well as other insects such as water beetles and dragonflies. The combination of little disturbed habitats and a prolific insect food supply has an additional effect. It leads to a rich and abundant birdlife during the summer months. In terms of numbers, golden plover, dunlin and greenshank lead the way but there are also redshanks, oystercatchers, ringed plovers, curlews and lapwings. The abundance of water also provides habitat for wildfowl. Teal, wigeon and red- breasted mergansers all breed in numbers of some importance but it is for rarer species that this area is especially important.

Complex pattern of peatland pools or 'dubh lochans', much favoured by breeding waders and wildfowl.

Peatland habitat, showing pools, moss carpets and moss hummocks..

The diminutive dwarf birch, growing in typical habitat amongst peatland pools.

Red-throated divers nest on the small peatland lochans and their black throated relatives frequent larger water bodies. This separation in habitat choice reflects their feeding preference. Black-throated divers are quite sedentary during the breeding season and generally obtain most of their food from their chosen nesting loch. Red-throats however, give priority to seclusion and nest on pool systems which may lack fish. Consequently they must commute regularly to favoured food sources, very often the sea. Another speciality is the common scoter. This duck, which has a very small national population and normally breeds in arctic latitudes, favours medium sized lochs which have clear water and firm mineral bases. They are less content on peaty pools with much emergent plant growth.

The North Scottish resident population of greylag geese numbers approximately 2000 birds and appears to be relatively stable. A high proportion of these birds nest within the peatlands, on the scattered pool systems. Because of the difficulties of undertaking census work on such a large and remote area of land the actual number of pairs involved is unknown but is likely to be about half of the total population. Unlike most birds which are totally dependent upon their ability to fly, waterfowl moult all their flight feathers simultaneously. Consequently all ducks and

Pair of red-throated divers on a peatland pool.

Dunlin (of the schinzii race) in full breeding plumage.

Golden plover.

geese become flightless for a period of about two weeks in late summer and their dependence upon secluded or large water bodies where they can take refuge during this vulnerable stage is self evident. Non-breeding greylag geese, or those which have lost clutches, will gather together in large flocks during June in order to moult. Virtually all those in Caithness and Sutherland without dependent young move to Loch Loyal in Sutherland for this reason. The remainder, unable to move because of their young, moult where they breed. The multitude of pools within the remote areas of peatland are perfectly safe.

Not all residents of the peatlands score highly on the attractiveness scale. Midges and cleggs have received a passing mention, but they deserve more than that, not because of their unattractiveness but because of the sheer misery they can instil into even the most hardened resident. They cannot be ignored and many an otherwise pleasant summer's day can be ruined by their incessant attention. Beware the clegs with green eyes, they sound as though they are powered by a petrol engine and they use knives!!

UPLAND

Deer forest

The transition from peatland or even valley side, into upland is often indistinct and indeed many upland areas contain extensive areas of peat dominated vegetation. The true peatlands are gentle in gradient and rarely occupy altitudes above 300m. Beyond this height and upon steeper gradients the soils become thinner, the amount of exposed rock increases and heather or upland grasses start to dominate the vegetation. This is the intermediate land, between the lowlands and the mountain summits. It is often remote and above all, on a grand scale. Contrary to its name, it is not a forest, although it may contain patches of woodland, usually relict birchwoods. It is, however, red deer country and deer stalking forms the major land use and form of management. In some parts, sheep may be grazed extensively but generally deer have exclusive use. Although this high country is now considered the usual home for deer herds, it should be remembered that the red deer is a native of woods and forests and it is only as a result of its adaptability that it has taken readily to the hills in response to a declining woodland resource over the centuries. Woodland remains the favoured habitat however, and given half a chance, most red deer will seek shelter amongst trees and will thrive. This is a major factor in the continued decline in the extent and quality of woodland remnants in the uplands.

In terms of vegetation, deer forests are diverse, for they cover a significant altitudinal range. They usually extend from the levels of upland rivers,

Deer forest landscape in Assynt.

usually the upper sections of the main rivers or tributaries of the same, up to the point at which the foothills become rugged mountains. This again is an indistinct division. The intervening areas may be primarily dry heather moorland, rocky or grassy in nature. At lower latitudes, from the southern Highlands southwards to northern England, this zone would be moorland, where red grouse would be numerous and form the principal landuse.

For much of the winter deer herds will stay within the lower sections of this land, often close to the rivers where grassy banks provide better feed. As growth resumes in spring they range wider up the hillsides until in summer deer are a rare sight at lower levels for they now utilise the upper slopes and rounded hill summits where they can feed on fresh growth of deer grass and escape the clegs at the same time.

This is the most suitable ground for birds such as ring ouzel which is one of the first migrants to return to the hills after their northward spring journeys from Africa. Merlins too enjoy the mixture of heather and grasslands where they catch large insects and small birds such as meadow pipits and skylarks. Golden plovers nest on the gentle slopes and common sandpipers feed along the burns and around riverside pools. Deer forest is also part of the main hunting range for golden eagles. They feed off a mixture of carrion, such as deer which failed to survive the winter and fresh prey such as mountain hare and grouse. Mountain hares undergo dramatic cycles in their population sizes and at times can be quite scarce before building up again to surprising abundance. They are not, however, as common in these northern parts as in equivalent habitats further south in Scotland.

Golden eagle.

Lochs and lochans

There are countless areas of standing water within Sutherland and these include many variations on a theme. In terms of size, there is everything from the small dubh lochan which can easily be stepped over, to Loch Shin, one of the longest stretches of water in the UK. The range of habitat types and altitudes is equally diverse. There are lochs at sea level as well as those at altitudes close to 3000 feet. Some are in soft substrates with abundant growths of sedges, reeds and rushes around the perimeter and others in hard, rocky locations where the water is clear and little vegetation grows. Some are artificial in origin, others natural but artificially retained or enlarged. The majority however are entirely natural.

It is in the zones of the hill country and deer forest that numerically at least, most of the lochs of the area are found. In the west in particular, the undulating terrain produced by the underlying Lewisian Gneiss rock leads to a landscape dotted with rocky outcrops and countless lochs. This is colloquially known, very descriptively, as the 'cnoc and lochan' landscape. Most of the lochs here are small and situated within depressions amongst the landform where surface drainage of the surrounding slopes is adequate to maintain the water level. Very often feeder and exit burns are indistinct and little more than ditches. More often though an intricate network of small burns links the lochs and eventually provides a vital artery to the main river system. By this means small fish can reach the hill lochans and often do well upon the rich aquatic insect life which thrives in the cool, clean waters. Vegetation

Cnoc and lochan landscape with Suilven in the background.

Loch Awe and Canisp beyond; a typical large Sutherland loch.

in these hill lochs is sparse, a characteristic of lochs with low nutrient inputs. Those species which are present are often of interest. Pondweeds, water millfoil, bladderworts and stoneworts are all present and the purple flowers of water lobelia may often ornament the water's edge. Small lochs within the Lewisian Gneiss country are often quiet for birds but surprise is one of their main qualities and it is often difficult to predict what might turn up. Common sandpipers can usually be relied upon and very often, so too can greenshank and teal. The larger lochs are often occupied by black- throated divers, especially where there are islands safe from land predators which can provide relatively safe nesting areas. It is rare for more than one pair to occupy a single loch but it is not unusual for favoured lochs to act as communal areas and groups of four, five or six may be found loafing together.

Montane
The highest altitudinal zone is that of the true mountains, of which there are many. There is no such thing as a typical mountain and all differ in their rock type and shape, degree of exposure and weathering. Consequently the variation in detail within the montane habitats is tremendous. The broadest of classifications could include those of rounded summits or those with craggy rock peaks.

The former usually have a weathered quartzite layer which leads to summits comprising an intricate mixture of scree slopes and low growing vegetation. A feature common to all mountain tops is the short growing season, high rainfall, low average temperatures and, above all, exposure to frequent and severe winds. These factors combine to eliminate all vegetation which requires anything other than a home amongst loose rock, where only meagre nutrient supplies are available from the thin, acidic soils.

The more rounded topped mountains have a sparse organic layer overlying much of the rock and this is generally covered by a montane heath, comprising wind clipped heather little more than one inch high, together with other dwarf shrubs of similar stature. These are commonly cowberry and bearberry or, more frequently at such altitude, the alpine version. Blaeberry is often replaced by the more hardy, northern billberry which is similar in most respects apart from a blue colouration to its rather more rounded leaves. Bare patches of dry peaty soil may also support the prostrate, trailing azalea or the rather curious high altitude, three jointed rush. The most obvious feature though, is not these flowering species but the extensive carpet of moss. This is the woolly fringe moss, the same as that which can grow as hummocks on peatland but here grows as a surface layer. The combination of all these plants produces an effect which is rather like walking upon an expensive carpet.

This rather special habitat is, not surprisingly, home to some rather special birds, as well as some more surprising residents. In summer, land of this type above 2500 feet is the breeding area for dotterel, a small plover which spends the winter months at low altitudes in Africa. The contrast cannot be greater. It is a quiet, unassuming bird which nests amongst the mosses and stones and relies on its camouflage for its safety. Apart from its unusual choice of breeding habitat it is equally unusual in its breeding habits. The female bird is the larger of the two and more boldly marked. Once it has laid the eggs it plays no further part in parental proceedings and abandons the male to incubate and rear the chicks. There are known instances where females may lay for more than one male, and not always on the same mountain.

Ben Loyal summit, looking west.

There are about 840 pairs of dotterel known to breed in Scotland, but the main breeding areas are in Scandinavia. In May, at the time of peak passage, parties of dotterel may stop on many mountains, and even low ground, but by June it is mainly only breeding individuals which remain.

Another bird of this high stony country is the ptarmigan, this time a resident which rarely ventures below 2500 feet at any time. In summer it raises its broods amongst the rocks in similar areas to the dotterel although it is freer in its choice and will also move happily into really rocky terrain. The adult birds take on a body colour almost inseparable form the background rock. It is often only the quiet purring sound or movement which gives them away. Unlike all of their other grouse relatives they are not especially jumpy birds and can be quite approachable. The onset of winter and frequent snows brings a change to the ptarmigan as they moult into their pure white winter plumage.

Snow buntings and shore larks are both high arctic species which nest in country not dissimilar from that of these mountain tops, albeit at lower altitudes. Consequently they are not out of place here and can sometimes be found during the summer months, although actual attempts to breed seem to be rare.

Birds at these altitudes have little choice of diet. In summer flies of various types are quite numerous and there is one species of crane fly which specialises in this upland terrain and on still days in summer can be quite numerous. Butterflies are not well represented but the strong flying migrant species such as red admiral can occur at altitudes of up to 3000 feet.

Female dotterel, a bird of high rounded hills.

Male ptarmigan, in transition from its white winter plumage into summer grey.

Another curiosity at these high levels is the water vole, normally associated with slow flowing meandering river systems. They do not occur on the summits themselves, but on the slopes. Amongst thicker vegetation, numerous drainage channels and small pools provide reasonably sheltered living accommodation for them. The population size is completely unknown and is likely to be small. The animals themselves are secretive and rarely seen, but the presence of holes and small tunnels linking to underground burns is a sure indication of their presence.

Wherever there are ground nesting birds or small voles, foxes are likely to occur. Control measures by land managers wishing to protect their

Rocky terrain of the moine mountains of Assynt.

game bird interests are usually less rigorous at these upper levels and foxes can often be seen in full daylight on the high summits.

The mountains of Sutherland and to a lesser extent, those of south Caithness, are diverse in structure and rock type, but many of the high peaks such as Ben Hope (927m), Ben Klibreck (961m), Ben Armine or Seana Braigh (927m) are composed of Moine rock. These imposing mountains have a different character from the more rounded quartzite-capped versions, even though they may lie in close proximity. Exposed rocks and crags are common features and vegetation is divided quite clearly into high altitude heathland, acid montane grassland or rock and ledge forms. It is the latter which differs from elsewhere and provides insights into the botanical richness of some of the areas protected from the ever searching mouths of red deer. Early spring, as soon as the snow layers start to recede, provides the first signs of colour as purple saxifrage flowers from its rocky footholds. Later, other cushion-like plants come into bloom but because of the exposure, are rarely flamboyant. Examples include moss campion, a dwarf relative of the pink campion, and mossy cyphel as well as a number of specialities.

Mountain avens and other arctic, alpine plants on high rocky ledges.

Ledges on cliff faces are usually rich in plant growth, especially where moisture seeps through the rock. Sometimes there are stunted trees, usually rowan or rarer members of the same family but often there are bright yellow globe flowers, water avens, meadowsweet, alpine lady's mantle, alpine thalictrum and early purple orchids. The list is long and each ledge is different. They represent examples of what can grow where plants are left undisturbed.

This rocky terrain is the real home of golden eagle, peregrine falcon and raven, all of which choose to nest on inaccessible rocky ledges where human disturbance can be avoided. These birds are all shy, eagles especially so and shun unwanted attention. Eagles start nesting early in the season and usually have eggs by the beginning of March. The nesting period is prolonged though and it may well be into August before the last of the young have departed the nest.

Conservation Issues

In common with most parts of the world, the conservation issues facing Sutherland are many, varied, complex and have no single cause. They tend to evolve over time and can affect different species, directly or indirectly as they do so. Some changes in bird populations can clearly be linked to human activity, others have causes which are not easily explained. There may be links to climate, habitat changes , fluctuations in predator numbers and/or other aspects which we can be tempted to say are outside our control. Others are more obviously influenced by land use policies. In either case, given sufficient political will, there is usually a solution.

As previous chapters have made clear, the extent and quality of blanket bog habitat and its important bird populations puts Sutherland on the world stage. The agricultural value of this land has always been low and much of its economic value has been linked to its sporting potential. In the mid 1980s one of the Government's land use priorities was to increase softwood timber production and attractive financial assistance was available to those prepared to set aside land for forestry. One component of the incentives was the offset of costs against tax. This, combined with low land prices was too tempting a carrot to be overlooked by those with large tax returns and land within the peatlands of Sutherland soon started to be ploughed and planted with conifers, driving resident species from their traditional breeding grounds. By the late 1980s this pattern was at epidemic proportions and vast tracts of high quality peatland and associated birdlife had been lost. There was much activity amongst conservation organisations to try to protect key areas and various plans and strategies were agreed. The publication *Birds, Bogs and Forestry* (Nature Conservancy Council 1987) sets out the issues in detail. However, it was the Chancellor of the Exchequer who resolved the problem, almost overnight in 1989, when tax incentives were withdrawn from planting costs.

New forestry planting on peatland during the 1980s.

Today, although pressure to plough and plant areas of peatland has receded and there is much greater awareness of their environmental value, we are left with the legacy of the 1980's policies. This land was never good forestry land and where trees have grown, windblow and tree diseases are becoming a common sight.

The value of peatland, not only as a valuable habitat for wildlife, but also as a carbon store is now recognized as important in countering the effects of climate change. In many cases poor, or failed forestry is being removed, with efforts to restore the bog surface. The RSPB is leading the way with a lot of this work, prioritising its efforts on areas of former richness.

Expansion of plantations on open country, which peaked in the 1980s, has led to a significant decline in the nationally important golden plover population. (See Whitfield 1997). Now nests and chicks of all ground nesting birds are under greater threat from foxes, crows and other predators which reside in the woodland and have easier access to neighbouring open ground to hunt.

The open landscapes of Sutherland now seem more secure against conversion to plantations, but they are not immune from other modern pressures. This land is exposed and windy and as a result has caught the attention of windfarm developers. This is yet another subject which polarises opinion. As a society we need to reduce dependency upon fossil fuels and renewable sources of energy need to be developed. However, we must not rush headlong into such development without understanding their influences and identifying appropriate locations. Development based upon opportunistic factors such as low land rentals and good wind capture data simply repeat the problems of the past. There is now convincing evidence that wind turbines situated on moorland habitat

Windfarm established on peatland habitat.

displace breeding bird populations, especially species such as golden plover. They also cause direct mortality to larger species, although quantifying this is challenging.

One of the earliest proposals for a windfarm in Sutherland was along the line of hills parallel to the coast south of Helmsdale. After much public angst this was rejected at public inquiry . There is no doubt that this proposal (of moderate scale by today's standards) would have been visually very prominent but there are now far more damaging developments in operation within the county. By 2016 there were 6 functioning windfarms in Sutherland, with a further 8 proposals in development, including a second attempt at one of the original Helmsdale sites. Some large scale proposals still include development on peatland. (A further 8 cases have been rejected as inappropriate).

Land use changes such as forestry or windfarms are dramatic and relatively sudden. Others are more insidious and do not attract the same amount of public interest. The loss of natural broadleaved woodland and species-rich grasslands are two examples. Changes in the quality and extent of these habitats are slow and comparisons between years not always obvious. Over decades, however the differences are striking and the extent of loss is now dramatic. There are two factors involved, one concerns the management of the population of a wild animal, red deer, and the other, domestic stock.

Since Victorian times red deer in Scotland have traditionally been managed by sporting estates. Under this type of management regime owners of large expanses of hill ground derive income from sportsmen who pay a premium to stalk deer, fish for salmon and shoot grouse. Incomes from this activity have been declining in some sectors over recent decades due to a mixture of circumstances. Factors include a dramatic decline and localisation in grouse numbers, a steady decline in salmon stocks and fragmentation of land holdings. As a consequence costs need to be cut and owners are looking to extract more from their land with less input.

The melting pot of influencing factors has led to increasing numbers of deer, particularly hinds. These animals are naturally woodland animals and require shelter from extreme weather conditions in winter. The increase in numbers leads to reduced woodland regeneration. This is not easy to detect in its early phases, but once mature trees start to disappear and there are no saplings to replace them the problem becomes critical. Income from stalking is primarily linked to stags. Shooting hinds costs land managers money and the incentive to reduce hind numbers is certainly not financial. Added to this is the perception that a relatively high proportion of hinds to stags is required in order to provide a consistent supply of stalkable stags. Advisors suggest a 1:1 ratio is all that is needed and many estates are carrying surplus numbers. Reduced tree cover within the straths of Sutherland reduces the amount of nutrients being returned to the soil, leading to reduced surface vegetation fewer insects and fewer birds and fewer fish in the rivers. Do we detect a vicious circle developing here!

Mature birchwood open to grazing by sheep and deer.

Ungrazed mature birchwood showing the presence of saplings and a lush ground flora.

There is also a social change taking place with fewer owners managing land as traditional sporting estates and more fragmentation of properties. This also introduces conflicts between land owners who have differing objectives. Some want to encourage natural woodland but cannot because of the deer from neighbouring land and so fencing becomes necessary. Fences then stop traditional seasonal movements of the animals and prevent the deer manager from managing the herd extensively. Pressure on the habitat therefore increases further. Fences also kill woodland grouse which have the unfortunate habit of flying into them. The complete absence of large herbivore grazing within fenced areas also produces unnaturally dense regeneration and ground vegetation. Woodland protection through fencing is very much a second best option.

Most often these land management pressures play themselves out locally and quietly. Occasionally the media and politicians become involved and a lot of other agendas creep in. At this stage it becomes difficult to focus on the real issue, habitat condition. There is a current risk of the debate on national deer numbers turning into an argument against sporting estate owners. It is of little consequence from the ecological perspective who owns land. How it is managed is of critical importance.

It is on these lines that politicians can assist, especially through agricultural support. The number of agricultural holdings with cattle has slowly declined and along with this change has come loss of grassland habitat. This may be due to simply abandoning areas of ground but is

Uncontrolled muirburn.

also linked to a reduction in hay making. Under-grazing or continuous grazing by sheep lead either to development of rush pasture with course grasses or a steady decline in species diversity. Numbers of breeding birds which utilise grassland habitats in Sutherland, such as curlew and lapwing are in a nose dive.

Is it not time that the annual April fire fest was addressed? Each year the story is the same, with large areas of hill, woodland or grassland damaged by uncontrolled but deliberate fires. A muirburn code was published in 2011 but much of the legislation relating to muirburn stems back to the Hill Farming Act of 1946 and it struggles to cover many of the scenarios played out today.

Some agricultural support has been so successful that secondary problems have arisen. Silage and intensive grass management has not only helped farmers manage stock but they have also unintentionally increased the winter survival of migrating geese, particularly greylags and pink-footed geese from Iceland. Each year approximately 360,000 pinkfeet and almost 90,000 greylags head for Britain in early autumn. In years gone by a few would stop in northern Scotland but the majority would continue south to eastern and southern Scotland, Norfolk and Lancashire. Due to the quality of winter grazing (and milder winters) many thousands now stay until spring. This is particularly the case around the Dornoch and Golspie areas of southeast Sutherland. Hunting of geese is commonplace in Iceland and each year between 30-40,000 greylags are shot. Despite this mortality the overall population has increased at a rate of about 12% over the last ten years. Much of this increase is due to improved winter survival. With large flocks feeding on agricultural land conflicts arise, especially in early spring as the geese help themselves to young grass earmarked for lambing time.

This chapter has concentrated on land use and land use changes but the marine environment has also been disturbed. Changes are not as easy to detect as with terrestrial habitats and it is often the animals linked to the marine environment which provide the indicators that all is not well. The decline in seabird populations since the start of the 21st century at the UK level is well documented and indicates significant regional variations in fortunes. Northern areas, traditionally some of the most numerically important, such as Shetland and Orkney seem to be faring worst. Here some species have failed to produce any young for a number of consecutive years. In Sutherland the pattern is less obvious. In general, the west coast colonies have performed better than the east coast and neither have shown such dramatic effects as further north. Food supply, or more accurately, availability of suitable prey close to breeding colonies, seems to be the root cause. Again at the national level, large species such as gannets are thriving and colonies are still expanding. These birds are generalists, able to feed on a range of fish species and crucially, are able to travel long distances in order to find food. It is the smaller birds, especially those like kittiwakes which feed at the surface, which have been hardest hit. These birds, together with auks, feed to a large degree upon sand eels. This highly nutritious fish is a very important component in the diet of the chicks of these species. Shoals of sand eels are no longer found in their traditional locations and breeding seabirds are faced with a choice of catching alternative, less nutritious prey or travelling longer distances. In the mid 2000s kittiwakes in the Moray Firth were feeding young largely on a diet of pipefish. This elongated fish has about as much nutritional value as a rubber band and most chicks were dying in the nest from starvation, even those with fish still in the beak. This phase seems to have passed, but sprats still outnumber sand eels in the diet. Many auks are travelling as far as the Tay to catch food, but they are unable to do this at a fast enough rate to sustain their young. Consequently recruitment is still low.

Why sand eels are now so patchily distributed is not fully understood but rising sea temperatures seem to be involved. As a consequence of warmer waters in northern Scotland the number of migrating shoals of mackerel is much higher. These are voracious predators of sand eels and could well be adding to the problems of nesting seabirds. Some scientists claim sea temperature changes are cyclical and by 2020 there should be a cooling effect leading to the possibility of recovery for breeding seabird populations. Only time will tell in this case. Meanwhile surviving adults need to contend with other pressures, such as marine rubbish, especially industrial plastic waste, as well as a new and, as yet unquantified threat from offshore renewable energy installations.

These preceding paragraphs may make for depressing reading as current trends for Sutherland's birds and wildlife in general seem to be negative. However, all is not gloom and there is much to be encouraged by. Barn owls have benefitted from a series of mild winters and are now established as a breeding species, particularly in the south-east. Red kites are breeding in Sutherland for the first time and without local reintroductions. Ospreys have increased noticeably and are now a regular sight around the East Sutherland coastline and elsewhere. Black throated divers seem to be at least stable and the conservation efforts of the RSPB to provide a network of floating nest platforms at some traditional nesting sites have locally increased productivity. Wintering flocks of wildfowl and waders on the intertidal sites are stable or in some cases increasing. Sanderling and grey plovers are now established wintering species in low, but ever increasing numbers.

Some of these changes are taking place unaided, others are a result of conservation activity. Sutherland has three Nature Reserves which are primarily for their ornithological interest. Handa Island is privately owned and run as a reserve by the Scottish Wildlife Trust, largely for its breeding seabird interest. Forsinard is a 20,000ha peatland reserve owned and managed by RSPB which covers some of the prime breeding areas for peatland birds. Loch Fleet is managed as a National Nature Reserve by Scottish Natural Heritage and the Scottish Wildlife Trust, in conjunction with the owners, Sutherland Estates.

In the past there were also National Nature Reserves at Inchnadamph, Alt a Mhuillin, Invernaver and Strathy Bog, all in northwest Sutherland, but these no longer meet the criteria for modern day reserves. They were largely created to protect sensitive habitats and encourage specialised management. Both of these aspects are now best delivered by other means and today National Nature Reserves concentrate more on visitor experiences.

The West Coast

Sandwood Bay.

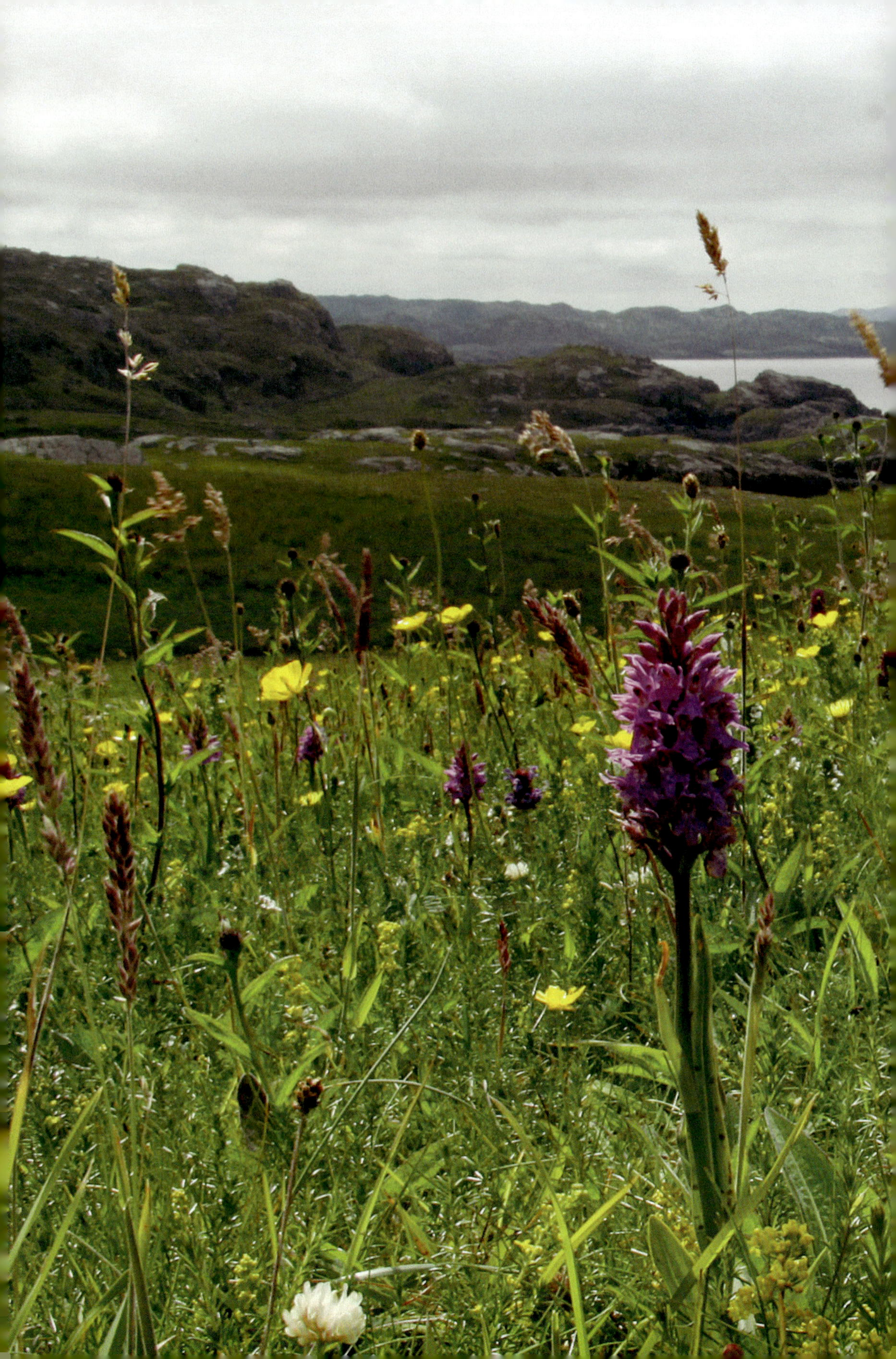

Flowering meadows on croft land near Kinlochbervie.

The North Coast

Looking east from Faraid Head, with Ben Loyal in the distance.

Strathy Point lighthouse.

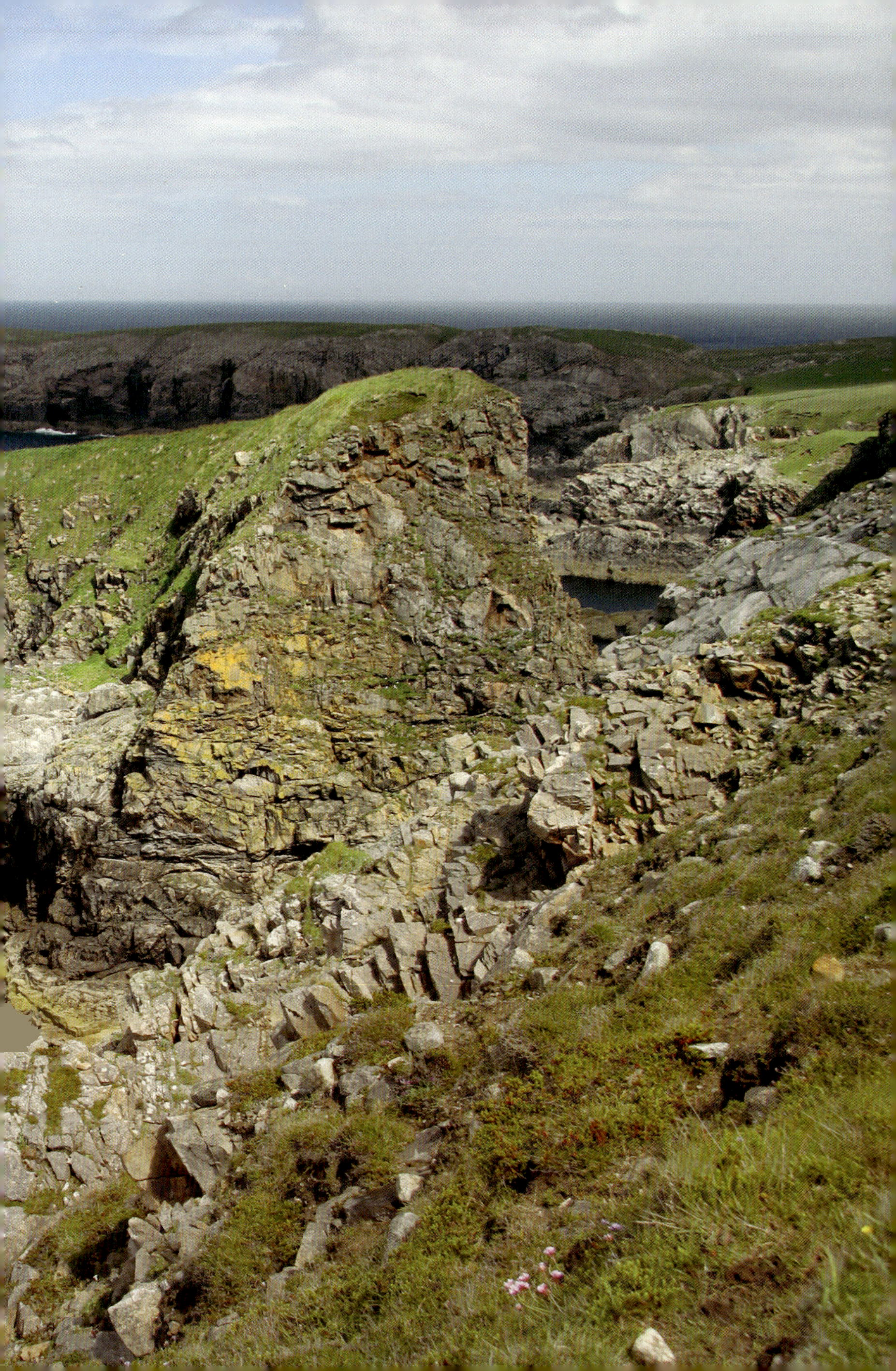

The East Coast

The outer Dornoch Firth.

Mound Alderwoods from the air.

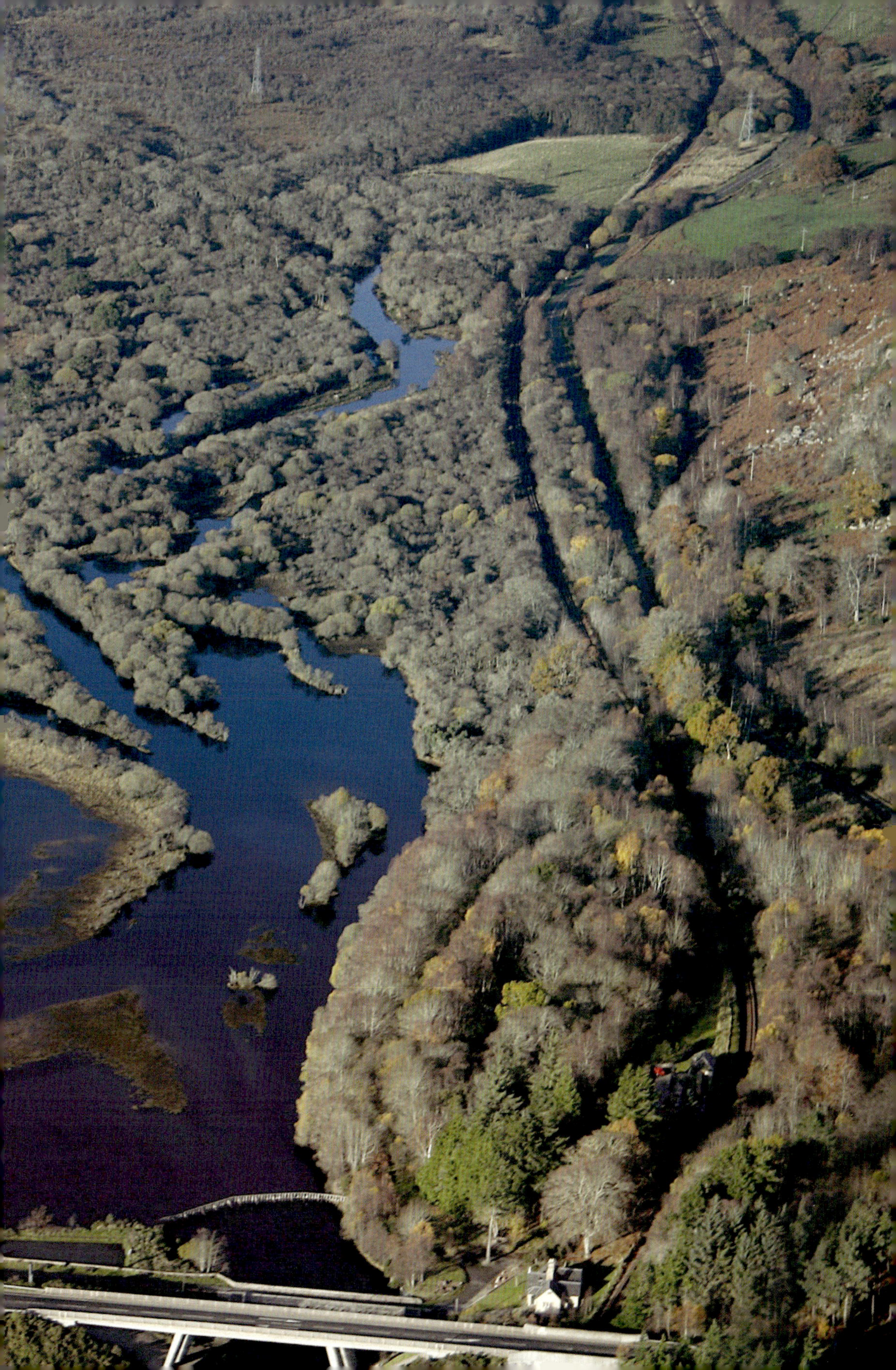

Rivers

The River Brora: one of many with origins within the peatlands.

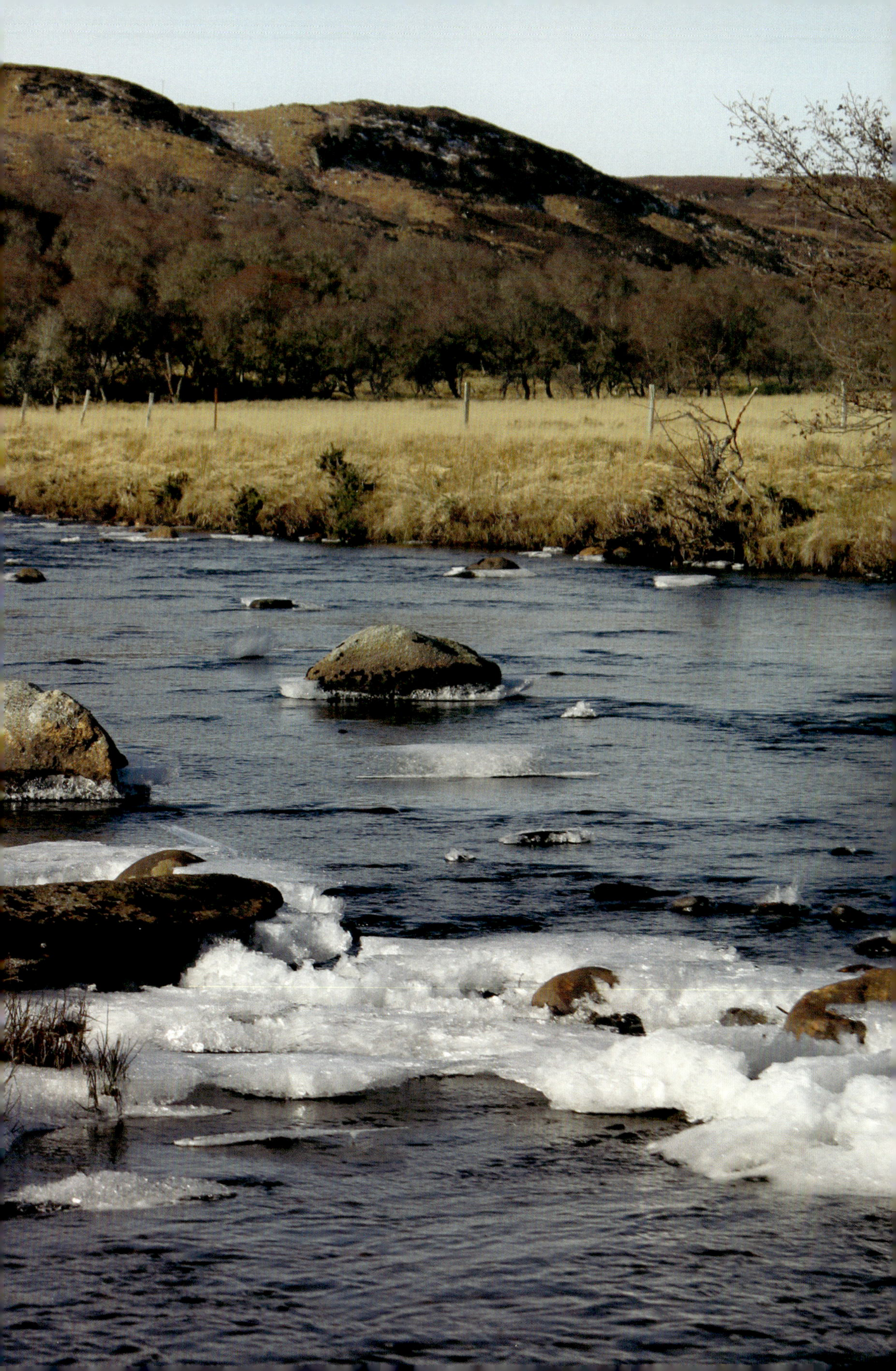

Winter on the River Carron.

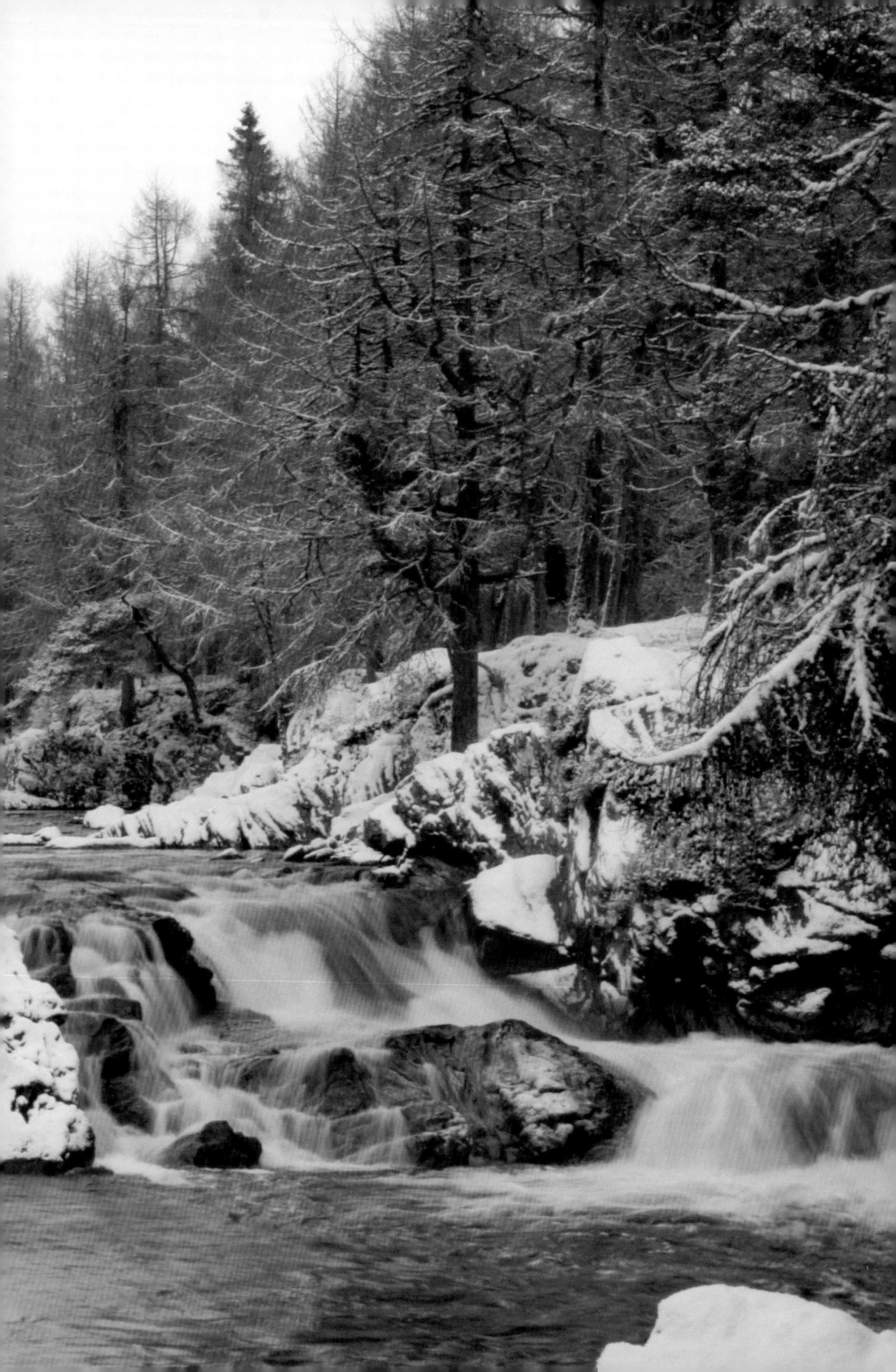

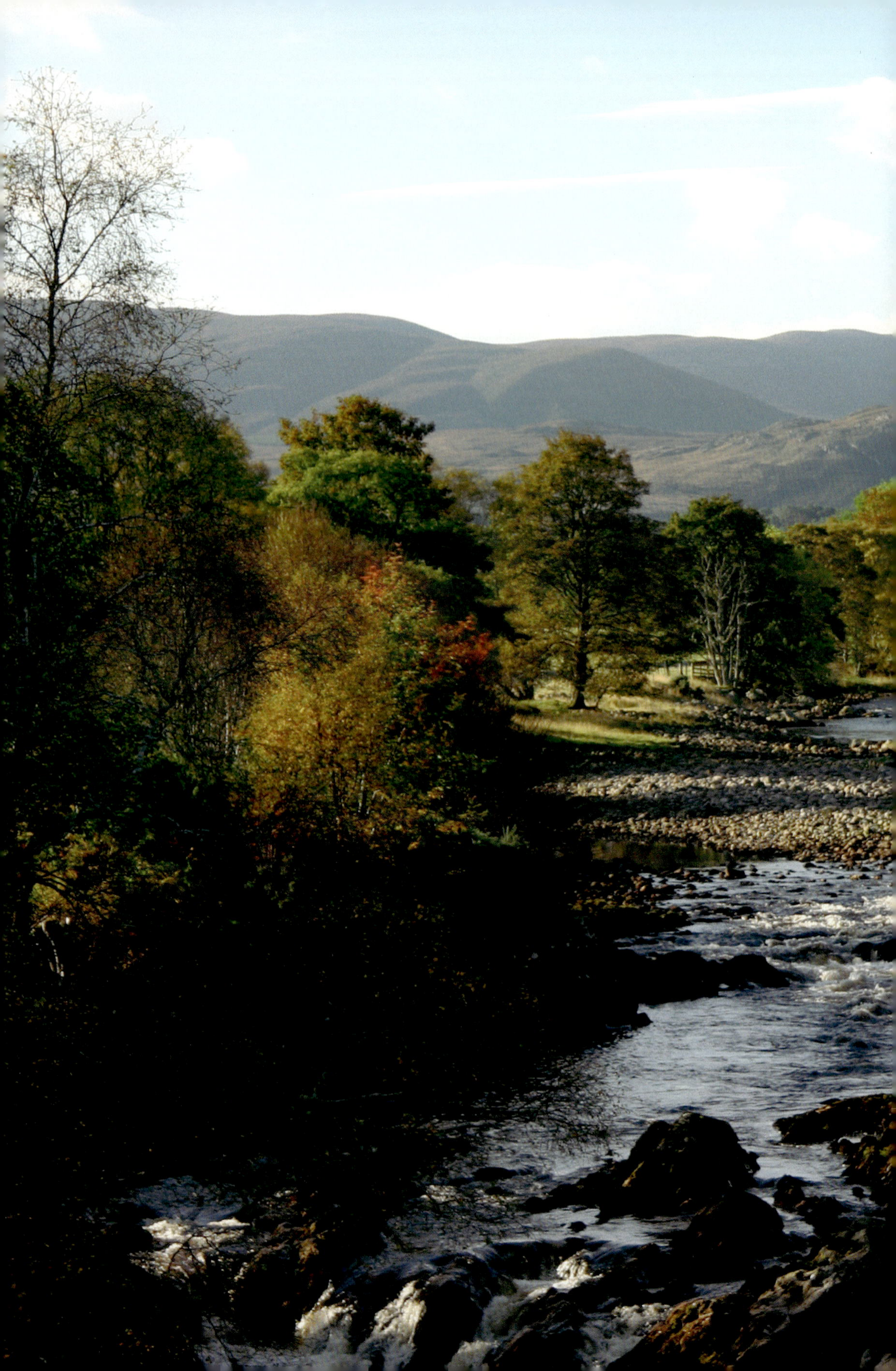

The River Carron in autumn.

Mountains

The home of ptarmigan; Quinag in wintery conditions.

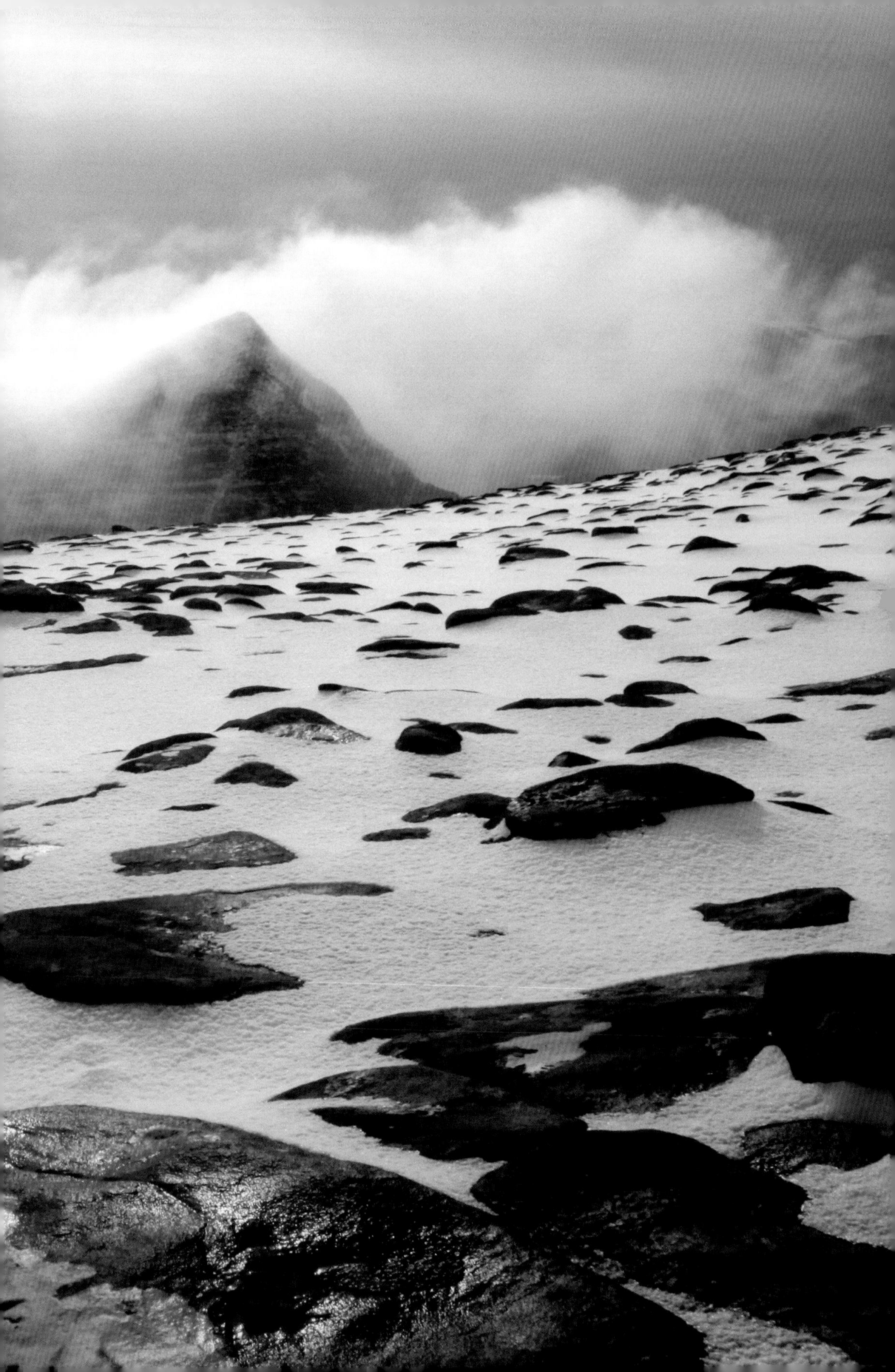

Sutherland breeding specialities

Temmink's stint.

Wood sandpiper. (Photo Dean MacAskill)

Sutherland winter visitors

Long-tailed ducks.

Whooper swans.

Sutherland resident

Lapwing in breeding plumage.

Birds in Sutherland

BREEDING SPECIALITIES

What defines a speciality may be open to debate as there are no set criteria. Most counties have their own rare breeding species and Sutherland is no exception. However some of these may be "one-offs" due to chance circumstances and may not be significant but others may be an indication of colonisation and the change of a species range. Whooper swan, great northern diver, ruff, whimbrel, fieldfare, brambling, common redpoll and snow bunting have all bred in the county but are not established as regular breeders. Redwings breed sporadically and their numbers are difficult to assess. These are all typical "Northern" or Arctic species and indicate the tendency for Sutherland to host occasional breeders at the southern end of their normal breeding range. Red kites and white-tailed eagles are now breeding in Sutherland but in numbers which are still in single figures. However their arrival is not unexpected due to the establishment and gradual expansion of these birds from re-introduction projects elsewhere. It can be expected that both will become routine breeding species in years to come, much like ospreys. This latter species did not start breeding in Sutherland until the 1970s but is now well distributed and is a common sight in several parts of the county, especially around the Dornoch Firth.

Sutherland's modest population of hen harrier assumes ever greater importance given their virtual extinction as breeding birds further south. Golden Eagle numbers seem to be stable, but this iconic species no longer breeds in the south-east, with the last pair displaced by an ill-sited windfarm. Honey Buzzard is probably a regular breeder in very small numbers in the south and goshawk is now more widespread. Hobby and Red-backed shrike may have bred and wryneck is a potential colonist. Temminck's stint is easily overlooked but a few known sites are occupied annually. Lesser whitethroat and garden warbler have become established, grasshopper warbler numbers have increased and singing icterine warblers have been heard in more than one recent year. Jays only appeared regularly in the county in 2005; breeding was not proven until 2015. Colonisation by common rosefinches, which seemed very likely twenty years ago, has not materialised.

Several eastern species are extending their range to the west quite rapidly and Sutherland is well-placed to provide some British 'firsts'. Thrush nightingale and Blyth's reed warbler are two such candidates, whilst two-barred crossbill has already bred just to the south in Easter Ross.

Although rarities are noteworthy, the most significant species are those which regularly occur in numbers representing a high proportion of the national population. Within this category golden plover, dunlin and red-throated diver deserve a mention, all of them peatland or moorland species which are sadly in steady decline. Greenshank too have retreated in range from some of their more peripheral breeding sites but wood sandpipers are showing small increases, although total numbers are still low. Black-throated diver and common scoter, both species of peatland lochs, are managing to maintain moderately stable populations, although the latter remains highly vulnerable.

MIGRATION

Although some notable movements of migrants have been recorded in the County in the past, it wasn't until the 1990s that intensive migration studies were carried out in the south-east. Coverage of other areas, with the exception of the extreme north-west (Durness), remains low.

The annual arrivals of summer and winter visitors, and passages of common transients are the most predictable examples of migration and the latter provides markers for the main flight lines used by other species. The most obvious of these is the NE/SW coastal strip, with its 'hostile' hinterland, from Helmsdale to Dornoch, which neatly links the Northern Isles with the Great Glen via the Inverness Firth. Equally important is the SE/NW fault running from the Dornoch Firth to Loch Laxford via Loch Shin, used by waders and other species (as well as thousands of pinkfeet) heading to and from Iceland.

River valleys funnel migrants that find themselves in the hinterland. N/S routes, such as Strath Halladale, are well used by birds coming to and from the Northern Isles. E/W straths also have their complement of travellers, particularly in 'drift' conditions when easterly winds can divert species from the Continent and Scandinavia across the North Sea in both spring and autumn. The north-east shoulder of Scotland intercepts most of these before they can reach Sutherland, but 'broad front' movements sometimes occur.

In spring, high pressure centred to the south causes some northbound migrants to overshoot their breeding grounds. This phenomenon accounts for a small annual harvest of 'southern' species, like black kite, white stork and perhaps even bee-eater, in most years. Some of these may be birds reorienting south from the Northern Isles, as a good proportion of them occur on or near the north coast.

The north coast also receives migrants crossing the Pentland Firth from Orkney after large autumn 'falls' in the Northern Isles. After resting and feeding, even some displaced night migrants choose to filter down through the mainland rather than wait for ideal night departure conditions. Large arrivals of waterfowl and waders from Iceland can be seen from northern headlands, such as Strathy Point, in autumn whilst Scandinavia probably provides the bulk of those passing down the south-east coast.

Irruptive species, like waxwing and crossbill, appear in numbers only when the food supply fails in their normal range. Movements are therefore not weather-related and can occur outside the main migration seasons.

Sudden cold snaps in winter can cause sizeable movements of geese, lapwings, skylarks and snow buntings, mainly from the large agricultural areas in nearby Caithness. It was probably a sudden freeze in southern Scandinavia that resulted in the unseasonal appearance of three stone curlews (the first for the County) on the south-east coast flight line in January 2009.

Whilst the migration of land and shore birds differs little from other parts of the mainland, Sutherland can stake a unique claim where seabirds are concerned. The 'mouth' of the Moray Firth extends for about a hundred miles from the north-east tip of the mainland (Duncansby Head in Caithness) to Fraserburgh, making it the largest seabird trap in Britain.

Seabirds enter this huge trap in three main ways: driven in from the northern North Sea by easterly gales; arriving from the Norwegian Sea after continuing to fly south-west when the Norway coast turns to the south-east near Alesund and coasting north from the southern North Sea and turning west at Fraserburgh in an attempt to reach the Atlantic.

Once in the trap, and finding their way blocked to the west, they exit the Firth to the north-east, passing the mouth of the Dornoch Firth and (often much closer inshore) the Brora 'bump' and Lothbeg Point, a few miles to the north-east.

Some identifiable individuals, such as an albino fulmar, have been 'tracked back' as far as the north Norfolk coast and a Fea's petrel, which passed Brora on 29 August 2003 was also seen at several Scottish headlands.

In ideal conditions (strong south-easterly winds ahead of an approaching depression), tens of thousands of the commoner seabirds, mainly auks, gannets, fulmars and kittiwakes, can pass in a single day and it is hardly surprising that a variety of uncommon and some supposedly rare species, like white-billed diver and Brünnich's guillemot, reward the committed seawatcher. Other rare Arctic species, such as ivory and Ross's gulls, are more likely to be seen here than in any other British location and even Pacific auklets are now a possibility (see p. 156). Even birdwatchers who are uncomfortable seawatching could hardly be unmoved by the sight of hundreds of little auks passing by, close inshore, in winter storms. Lack of coverage at the main watchpoints, it seems, is the main limiting factor in revealing the true extent of seabird movements.

In the north, the Pentland Firth offers the first marine access to the Atlantic and is used routinely in autumn by skuas and Manx and sooty shearwaters, sometimes interspersed with the rarer 'southern' shearwaters, and smaller numbers of petrels, phalaropes and Sabine's gulls. Strathy Point is ideally situated to view such movements and was the focal point for an unprecedented passage of over 1400 great shearwaters on 8 September 2007 (see p. 104).

The Point of Stoer in the south-west of Sutherland is a good location to monitor passage through the eastern Minch, although birds tend to be further offshore and the viewing points are too high for ideal observation. Feeding petrels can be fairly common in early autumn and it is also a reliable place to see cetaceans.

Climate change will undoubtedly influence the timing, scale and variety of seabird movements. Sooty and Balearic shearwaters, for example, are already occurring later in the autumn. As sea temperatures rise, species such as Fea's and Wilson petrels are likely to venture routinely further north. More extreme local weather will drive greater numbers of seabirds inshore and the higher winds predicted for the Arctic regions will drive birds south. These are now more likely to include rarities from both the North Pacific and eastern Siberia, such as the harlequin ducks now appearing with greater regularity in eastern Scotland. There has never been a more exciting time for British seawatchers, as just about anything is possible.

KEY BIRDING SITES

Sutherland is blessed with such a wealth and variety of 'wild' land, it would be possible to find interesting birds almost anywhere. Many of these areas are remote, in difficult terrain, and this Section therefore concentrates on the more accessible localities where northern specialities are likely to be seen or which are known to attract a wide variety of species. Good vantage points for watching coastal movements are also included.

Dornoch Point and the outer Dornoch Firth

The long, sandy spit extending south from Dornoch has wintering snow buntings. A high tide wader roost can contain scarce migrants and the occasional rarity, most notably greater sand plover, Kentish plover and Baird's sandpiper in recent years. Passerine migrants have included Richard's pipit and Siberian chiffchaff. Black terns are regular offshore in early autumn. The saltings and tidal flats inside the Point hold large numbers of common waders and wildfowl, with sometimes hundreds of pintail in the autumn. It is the best Sutherland site for wintering grey plovers, when hen harrier and short-eared owl may also occur. Footpath access to the Point from Dornoch airfield and to the tidal flats from the Dornoch/Cuthill minor road, near the A9.

Migdale Woods

A mixed pine and deciduous woodland on the shore of Loch Migdale managed by the Woodland Trust, just to the north of the inner Dornoch Firth. Best known for its singing wood warblers in spring but rich in other species including tree pipit and redstart. Red squirrel still occurs here. Red-throated diver breeds on the loch. Frequent sightings of osprey and peregrine and a chance of honey buzzard. Minor road access to loch-side footpath from both Spinningdale in the east and Bonar Bridge in the west.

Kyle of Sutherland and Invershin

The fields below Bonar Bridge have wintering greylags which are sometimes accompanied by rarer species like white-fronts. Canada geese are resident here. To the west, the Kyle broadens and hosts good numbers of goldeneye, tufted duck and goosander in winter. The reedbed at Invershin is always worth a scan from the lay-by opposite at any season – perfect for marsh harrier, although it was a juvenile Montagu's Harrier that was found dead near here in October 2002. Main road access.

Embo/Dornoch

In winter, this stretch of coastline attracts the largest flocks of sea duck (common and velvet scoter, eider, long-tailed duck), all three divers, up to a hundred slavonian grebes and the occasional great crested and red-necked grebes. The exact location of the birds is partly tide and wind dependent, so it may be necessary to scan both from the Embo caravan site and the Dornoch beach car park. Little gull is regular here and the rocks often have waders like purple sandpiper and knot. Access by minor roads.

Loch Fleet and The Mound
This superb, unspoilt tidal inlet, with only a narrow access to the sea at Littleferry, attracts several thousand birds at peak times including most of the common waders, shelduck, wigeon, eider and red-breasted merganser, whilst flocks of greylag geese often rest here. More than 20 greenshanks now overwinter and fishing ospreys are a frequent site in summer. Strangely, the number of uncommon migrants and rarities seen is very small given the regular coverage it receives, but greater yellowlegs, black duck and black stork are notable exceptions. The Mound alderwoods may yield singing redwing in spring and spotted crake has also been heard here. Excellent viewing from minor road along southern shore and the car park at The Mound.

Strath Carnaig and Loch Buie
Inland from Loch Fleet a minor road climbs up through birchwoods into open moorland, reaching Loch Buie after 6 miles. Black-throated diver is regular here in summer, sharing the loch with a few breeding greylags and common sandpipers. The area is raptor-rich, with osprey, hen harrier, merlin and peregrine all likely. Following a large winter influx of common redpolls, birds have nested in the plantation on the south ridge in several subsequent years and a male stonechat of the white-rumped Continental race *rubicola* also bred here, paired with a local female.

Big Burn, Golspie
Dipper is the main attraction here, but the varied woodland hosts a good selection of common species, including chiffchaff and blackcap. This was the first place jay was proved to breed after its recent arrival in the County and green woodpecker (another likely colonist) has been reported there. Network of well maintained footpaths accessible from main road in Golspie (A9) and minor road on northern side.

Brora

Best known pre-1990 for its regular king eider, intensive watching since then has shown Brora and its surrounds to be the Sutherland hot-spot for rare seabirds and other migrants. The reasons are complex, but the 'Brora bump' is the first promontory seabirds trapped in the Moray Firth have to negotiate as they head back out to the east, which also intercepts coasting birds heading SW in autumn. The Clynelish valley behind the village is the natural short-cut for migrants heading SW down the coastal flight line and some linger in the 'softer' habitats they encounter after crossing the harsher terrain in Caithness. The headland, which has an extensive area of rocks and sand for gulls to loaf and waders to feed at low tide, is also due north of the tip of Tarbatness for those migrating by compass. Divers and sea ducks congregate in Kintradwell Bay, just to the NE, and can be scanned from the dunes bordering the golf course.

The car park at Lower Brora is the best place to monitor passing seabirds and coastal migration. Time spent there is rarely wasted in any conditions, but in strong south-easterlies, particularly in autumn, the volume and variety of passage can be breathtaking. A 'classic' movement can produce rare shearwaters, storm and leach's petrels, grey phalaropes, all four skuas, little auks and perhaps a Sabine's gull or two. National rarities like white-billed diver and Brünnich's guillemot have proved to be quite regular, the latter mainly in November. Iceland and glaucous gulls are regular in winter and spring and a few Mediterranean gulls are now joining the overwintering little gulls. There have been several sightings of ivory and Ross's gulls.

The number of rare migrants recorded in the Brora area is staggering, ranging from harlequin duck and south polar skua to booted warbler and yellow-breasted bunting, via broad-billed sandpiper and kentish plover. Even the celebrated 2009 sandhill crane checked in briefly! But no king eiders for a while.

Strath Brora

A delightful minor road leaves Brora village by the clock tower, passes Clynelish Moss on the right and the six mile long Loch Brora on the left, before climbing and turning to the south-west towards Rogart. The loch sometimes has feeding divers in summer and the surrounding woodland (mainly birch) has its complement of songsters which can now include garden warbler and lesser whitethroat. There is a fulmar colony on the spectacular Carroll Rock which deters breeding raptors, but there are plenty in the area, including hen harrier, goshawk, the occasional white-tailed eagle, and golden eagle in winter. Spotted flycatcher and whinchat are still fairly common, both common and Scottish crossbills occur and twites breed sparingly in the flanking hills. Beyond Sciberscross, the river turns back to the west into less accessible terrain, but the road continues on past the reed-lined Rhilochan, where water rail almost certainly breeds, sedge warblers are common and long-eared owls hunt in the late evening daylight. Black grouse and short-eared owls are regularly seen in the West Langwell area.

Loch Shin
This 20 mile long waterway forms part of the NW fault flight line for birds migrating to and from Iceland, like pinkfeet and whoopers. At its south eastern limit at Lairg, Little Loch Shin is good for diving ducks including the occasional smew in winter, and black-throated divers are regular visitors in summer. Just north-west of Lairg the Colaboll inlet at the mouth of the River Tirry has a rather ill-sited public bird hide and, depending on water levels, the pools can be good for migrant waders like black-tailed godwit. The 2001 Argyll snowy egret stopped off here, having presumably arrived in Britain via Iceland, and it is a likely site for other vagrants. The viewing from the road on the northern side is sometimes more productive. The road then follows the loch throughout its length and there are many points suitable for scanning. A stop at the River Fiag crossing can be worthwhile. Black grouse are most likely in plantations towards the northern end.

Forsinard
This RSPB reserve is worth a visit for the peatland interpretation, both in the reserve centre (in the old railway station, right beside the road) and on the nearby short boardwalk. The main areas of bird interest are well away from the reserve centre. So do not expect to see 'Flows' specialities like common scoter unless you are able to organise guided access and are prepared for a long flog!

Strathy Point
Despite the seabird riches at Brora and Lothbeg Point described above, for the ultimate seawatching experience there is nothing to beat Strathy Point in a strong westerly in autumn with birds battling through the Pentland Firth over huge Atlantic 'rollers'! The less agile can view this from the shelter of the lighthouse buildings (a half mile walk from the car park), but there is a tricky route out across a causeway of slanting rocks to the outer point, where an east-facing ledge provides sheltered, dry viewing. There is a regular passage of Manx and sooty shearwaters, with hundreds of the latter in good years. Check the passing great northerns for a white-billed diver. The few Balearic shearwaters are usually close in and sometimes stop to feed amongst fulmars in the broken water off the point. Great and Cory's shearwaters also occur (over 1400 of the former on 8 Sep 07). Storm petrels are most numerous in August (max. 800); Leach's more likely in September (max. 51). Pomarine and long-tailed skuas pass in variable numbers and multiple Sabine's gulls are not infrequent. Factor in the thousands of gannets, fulmars, kittiwakes and auks, the arrival of waders and wildfowl from the north-west, possibly including some Greenland white-fronts, plus barnacles and pale-bellied brents arriving from the north-east, and you may start to get a feel for the magnitude of a 'big day' here.

There is still much to see off the Point in more benign conditions and at other seasons. Risso's dolphins and minke whales are then more visible, with other rarer cetaceans possible. The Scottish primrose is found here on the short cliff-top sward. Coasting oddities have included smew, garganey and purple heron. A hobby was seen coming in from Hoy and passerine migrants have included ring ouzel and Richard's pipit, with snow buntings towards winter. Rather a good place for a bird observatory!

Kyle of Tongue and Portvasgo
The long finger of the Kyle extends five miles inland amongst spectacular scenery below Ben Loyal and is ringed by a minor road. A causeway with car parks allows good viewing of an arctic tern colony and the outer waters. The sandy tidal flats do not host large numbers of ducks or waders, but greenshanks are fairly common (April high tide roost of 32 near Kinloch). Great northern diver is resident in winter and

wildfowl can include brents and whoopers, with one Bewick's swan seen recently. Sutherland's first avocets since the recent expansion of its British breeding range occurred here in Jan/Feb 2016. This is golden eagle country, so keep your eyes on the sky! The Tongue area has had some interesting migrants: a singing rustic bunting in late spring; American wigeon and pectoral sandpiper in autumn to name but three.

A local ornithologist, Martin Scott, discovered that seabirds driven into the mouth of the Kyle in westerly gales exit it, very close in, past Portvasgo – a tiny inlet to the west of the mouth which is reached from the hamlet by a short grassy track. Although viewing conditions are not ideal, it is possible to scope from the car. Martin saw 9 Sabine's gulls (6 adults) here on 22 Aug 2003 (when none from Strathy Point) and there have subsequently been sightings of both grey and red-necked phalaropes and Leach's petrels as well as of the other Pentland Firth regulars.

Loch Eriboll

Deeper water gives this sea loch a different character from the Kyles of Tongue and Durness, to the east and west respectively, and many more birds. The three divers can all be seen here in winter along with slavonian grebes, mergansers, long-tailed ducks and auks. The latter may include little auks after winter storms. Watching them feeding is a very different experience from counting storm-blown birds passing offshore. The road on the eastern side provides good views over the inner loch.

Durness

This beautiful area in the extreme north-west is so different topographically and geologically from the surrounding areas. It has many of the attributes of an island and indeed its birdlife resembles that of the Northern Isles. Its oligotrophic lochs have breeding tufted ducks and ring-necked duck has occurred several times in spring. The water levels in the marsh at Balnakeil are now maintained at a higher level, encouraging an expansion of the boundary dyke reedbeds and providing breeding habitat for wildfowl, moorhen and perhaps water rail. Around a thousand barnacle geese feed here in winter and other geese, including Greenland white-front and a few taiga beans may occur. A small population of corncrakes is also being assisted by conservation measures.

Balnakeil Bay has great northern divers and long-tailed ducks in winter plus a few black guillemots. Faraidh Head offers possibilities for watching offshore seabird movements, but is a long walk from the beach car park in wild weather and can be off-limits during aircraft bombing activity. The cliff-top to the west, beyond the village refuse disposal point, is an alternative vantage point, accessible by car, although rather high.

Passerine migrants can find shelter and feeding in the gardens of the craft village or the pine wood on the golf course above Loch Borralie. Like the northern isles it resembles, Durness has a long list of rarities to its name, most notably Daurian starling, desert wheatear, woodchat shrike, Pallas's warbler and yellow-breasted bunting. Almost anything could turn up here, and probably will!

Clo Mor and Cape Wrath
The cliffs here form one of the largest and most extensive seabird colonies in Sutherland. Most notable are the puffins, which number tens of thousands, but kittiwakes, guillemots, razorbills and shags are also abundant. Many of the large auk colonies are within boulder fields at the foot of Clo Mor, whilst the upper, well vegetated slopes are the principal location for the puffins. The scenery here is dramatic and large scale. It lies within the MOD air to ground military training area and so access is only at non-active times. The passenger ferry from Keodale is the most practical approach.

Handa island
The Scottish Wildlife Trust run reserve at Handa is accessible via passenger ferry from Tarbet during the summer months and makes for a very enjoyable trip. Cliff nesting seabirds are the main attraction, especially fulmars, kittiwakes and guillemots but a few puffins are now returning to the island since the eradication of rats. The interior of the island is the only regular breeding location in the County for both great and arctic skuas. In some years arctic terns nest on the shores. Wheatears are especially abundant and occasional uncommon migrants may be found during the main periods of spring and autumn movements.

Sutherland Birdlife in Watercolours

The following pages show watercolour illustrations inspired by some of the birdlife found within Sutherland at different times of year.

Ringed plovers.

Wintering Waders

Curlews.

Knot.

Breeding waders of the peatlands

Greenshank.

Golden plovers.

Breeding seabirds

Puffins.

Guillemots.

Kittiwakes.

Fulmars.

Waterfowl

Eider flock.

Greylag geese.

Pink-footed geese.

Sutherland resident species

Black grouse.

Ringed plover with chicks.

113

Golden eagle

Redwings feeding on autumn rowan berries.

Peregrine falcon.

Sutherland winter visitor

Long-tailed ducks in winter plumage.

Annotated Species List

By the end of 1996, when 'The Birds of Sutherland' was in preparation, 284 species had been recorded in the County. Of these, 130 were regular breeders with a further 20 which had bred at least once in the last century. In the twenty years up to the end of 2016 a further 43 rare migrants or vagrants had been added to the list and seven new species (whooper swan, shoveler, garganey, great northern diver, lesser whitethroat, jay and common redpoll) have been proved to breed, whilst honey buzzard and water rail have almost certainly done so. To these must now be added red kite, a recent re-introduction, along with the welcome return of white-tailed eagle.

The recognised order of species has changed as DNA analysis throws new light upon lineage and inter-specific relationships. The list within the subsequent pages follows that now adopted by the British Ornithological Union. Recent 'splits' of closely related forms into separate species, with others likely in the near future, also affects the species total for Sutherland, which now stands at 328.

'The Birds of Sutherland' (1997) provided a detailed assessment of the status of each species in the County. The following accounts concentrate on any changes which have occurred in the last twenty years. Individual records are included only in the case of birds new to the County, extreme rarities or sightings of an exceptional nature, such as a major influx or movement, or an unusually early or late date.

Suspicion of single observer records unsupported by photographs by the Scottish and British rarities committees has resulted in the non-submission of many interesting sightings in recent years. Where there is absolutely no doubt about the identification, they are included here to ensure future researchers obtain a proper picture of occurrences in the far north, some of which may provide early indications of a developing trend.

Bean geese of *rossicus* race with a greylag at Loch Fleet.

Mute Swan
Cygnus olor
Breeding resident in south and south-east. Status unchanged.

Bewick's Swan
Cygnus columbianus
Two recent records: adult arrived from W, Loch Brora, 28 Mar 99; one Kyle of Tongue, late Oct 01. First confirmed sightings.

Whooper Swan
Cygnus cygnus
Passage migrant and winter visitor. Summering individuals frequent, with confirmed breeding at one site in 2014 (4 young fledged) and 2015 (3 young). Adults were present again in 2016 but breeding was not attempted.

Black Swan
Cygnus atratus
Escape. A single bird came in off the sea at Embo with a family of 6 whooper swans on the 29 Oct 14.

Taiga Bean Goose
Anser (fabalis) fabalis
Uncommon passage migrant and winter visitor, Oct – Apr, but increased numbers in recent years. Main arrivals Oct/Nov and Jan, with 24 flying W off Durness on 15 Jan 16.

Tundra Bean Goose
Anser (fabalis) rossicus
Rare passage migrant and winter visitor. Earliest, one Brora 16 Sep 07; max. 3 Brora (later Loch Brora) 22 Oct 08. An adult with 2 juveniles at Loch Fleet during a large passage of greylags and pinkfeet, on 12 Nov 11. *See photo left.*

An exhausted bird of the very rare pink-legged form *'neglectus'* (known as 'Sushkin's Goose') arrived at Brora in a SW gale on 7 Nov 05. This form has occurred at least once before in Scotland. Baxter & Rintoul (1953), Vaurie (1959).

Pink-footed Goose
Anser brachyrhynchus
Abundant passage migrant and (increasing) winter resident in south, Sep – May. Occasional summering individuals probably injured birds. Earliest arrival 06 Sept 12 (17 in off the sea at Dornoch).

White-fronted Goose
Anser albifrons
The nominate form is an uncommon passage migrant and winter visitor of erratic occurrence, maxima 9 Loth 21 Feb 04, 7 in from east, Brora 26 Nov 08 (accompanied by a Taiga Bean), 28 Loch Fleet 23 Nov 11 & 30 Dornoch 28 Feb 12.

Greenland White-fronted Goose
Anser (albifrons) flavirostris
Uncommon passage migrant and winter visitor. Main arrival Oct, earliest 7 flying W, Strathy Point 3 Oct 03. With several hundreds wintering in Caithness, most 'White-fronts' occurring in Sutherland belong to this distinctive race, which probably deserves specific status. On 29 Oct 07, 9 birds which passed Strathy Point NW/SE were no doubt Caithness arrivals but the 8 which flew W were more likely to have been Islay or Irish winterers.

Greylag Goose
Anser anser
Common and increasing breeding resident. Icelandic breeders also common on passage and in winter.

Bar-headed Goose
Anser indicus
Escapes or wanderers from feral European populations occur occasionally.

Ross's Goose
Chen rossii
Probable escape. Adult of unknown origin at Loch Fleet on 19 Apr 11 and at Golspie on 20 Apr 11.

Snow Goose
Anser caerulescens
One (of suspect origin) at Durness on 8 May 98. A pale-morph there from 19 Nov – 1 Dec 06 and a blue-morph at Loch Fleet 29 Sept – 1 Nov 09 are more likely to have been genuine vagrants.

Canada Goose
Branta canadensis
A small breeding population in the south-east is occasionally augmented by overshooters from the Beauly Firth 'moult migration', e.g. 8 at Rhilochan on 12 June 02. The species is now breeding in Iceland and one was with pink-feet flying N past Stoerhead on 18 Apr 03.

Lesser Canada Goose
Branta hutchinsii
Probable escape. Two birds showing the characteristics of this form were with moulting greylags on Loch Loyal in July 03.

Barnacle Goose
Branta leucopsis
Passage migrant and winter visitor, Sep – Apr. About a thousand birds winter in the north-west, arriving in Oct (529 flew W past Strathy Point on 3 Oct 03). Ten early migrants flying SW past Brora on 10 Sep 00 were probably heading for Caerlaverock, like most

Male shelduck.

of the later flights along the south-east coast, e.g. 378 on 24 Sep 06.

Brent Goose
Branta bernicla
The dark-bellied nominate form from the north-east is an uncommon passage migrant, mainly in Apr and Sept/Oct and usually occurring singly.

The pale-bellied form *hrota* from Greenland and Svalbard is a regular passage migrant in small numbers, mainly Sept/Oct, but two flew W past Strathy Point on 30 Aug 04 and a record 171 flew W there on 27 Sep 08. Odd birds overwinter, sometimes with barnacles.

Red-breasted Goose
Branta ruficollis
Probable escape. An adult accompanied greylags in the Brora area from 24 – 29 Apr 12.

Shelduck
Tadorna tadorna
Common breeding resident, mainly in south east. Passage birds occur on all coasts; occasionally ventures inland. *See photo above.*

Drake wigeon grazing on saltmarsh.

Ruddy Shelduck
Tadorna ferruginea
There have been no sightings since the influx in the late nineteenth century. Any new records would almost certainly be of feral birds, which have become increasingly common in Europe.

Mandarin
Aix galericulata
Occasional visitor, probably from the Highland population not far to the south, but no evidence yet of breeding.

Wigeon
Anas penelope
Sparsely distributed breeder; abundant passage migrant and winter resident. *See photo above.*

American Wigeon
Anas americana
Vagrant, A male (photographed), Kyle of Tongue, 11 May 03, male at Dornoch Point throughout Jan 15.

Gadwall
Anas strepera
Formerly a rare, sporadic breeder but now breeding regularly in the north-west. Probably as a result of an increase in the Iceland breeding population, now occurring more regularly as a passage migrant, particularly in the north. A record total of 34 passed Strathy Point on 29 Oct 07.

Teal
Anas crecca
Widespread breeder; abundant passage migrant and winter resident.

Green-winged Teal
Anas carolinensis
Vagrant. Last recorded in April/May 1995.

Mallard
Anas platyrhynchos
Widespread and abundant resident.

Black Duck
Anas rubripes
An adult female at Loch Fleet from 26 Oct 99 well into 2000 was the first County record. Possibly the same there from 3 Jan – 14 Feb 01.

Pintail
Anas acuta
No recent proof of breeding. Passage migrant, mainly in Sep/Oct, when numbers in the Dornoch Firth regularly reach three figures. Maximum 400 there on 19 Oct 08. *See photo below.*

Pintails are a regular sight amongst wigeon flocks in the Dornoch Firth in autumn.

Garganey; a sporadic breeder in Sutherland.

Garganey
Anas querquedula
A pair with 3 chicks at a site near the west coast on 8 June 97 was the first proof of breeding. Since then, a male at the same site in May 00 and probable breeding in the north-west in 2015, with 2 or 3 birds present at one site in Sep. Otherwise, a rare migrant. A juvenile at Durness in late Sep/early Oct 98 and one flying W off Strathy Point on 27 Aug 05. *See photo above.*

Blue-winged Teal
Anas discors
There have been no further reports of this American vagrant since the sightings of a male at the Mound in April 1993.

Shoveler
Anas clypeata
A pair was present on Handa in mid-April 97 and breeding was proved there in 2003. The species now also breeds at Balnakeil marsh, Durness. Otherwise, an uncommon migrant, max. 12 flying SW off Brora on 9 Aug 00.

Red-crested Pochard
Netta rufina
Vagrant or possible escape. No sightings since the male at Durness in May 1985.

Pochard
Aythya ferina
Has bred, but no recent reports. This species does not favour the oligotrophic lochs found in most of Sutherland and is a surprisingly uncommon coastal migrant.

Ring-necked Duck
Aythya collaris
This American species occurs on the lochs at Durness with some regularity, mainly in Mar/Apr. Also a male at Loch Evelix between the 12 Feb & 09 Mar 10 and presumably the same bird from 07 Nov 10 to 02 Feb 11.

Tufted Duck
Aythya fuligula
The two strongholds of this species are the Kyle of Sutherland, where several hundred overwinter, and the Durness lochs, where there is a resident breeding population of about twenty pairs. Uncommon migrant elsewhere.

Scaup
Aythya marila
Has bred, and the occasional sighting of males in summer in the north suggest it may be overlooked. Arrivals of wintering birds noted at Brora and Strathy Point in Sep/Oct. The only regular wintering flock is in the Dornoch Firth, although these birds are most often on the Easter Ross side.

Lesser Scaup
Aythya affinis
A large increase in the American population of this species accounts for its recent elevation from extreme rarity to regular visitor to British waters. The only Sutherland records so far are from the Brora rivermouth: 2 juveniles on 30 Oct 07, a female on 22 May 08 and 2 females on 23 Oct 08 – possibly all relating to just two individuals.

Common Eider
Somateria mollissima
Although still a fairly common breeding resident, numbers of this species have declined drastically on both the Moray and Pentland Firth coasts, possibly as a result of rising sea temperatures affecting their food supply. A male showing the characteristics of the northern race *borealis* has wintered off Embo every year since 2009. *See photo overleaf.*

Drake eider duck.

King Eider
Somateria spectabilis
Since the spate of records in the south-east in the 1980s and early 90s, the only record is of a male off Brora, which landed well out to sea in Kintradwell Bay on 9 Nov 08.

Harlequin Duck
Histrionicus histrionicus
There have been three sightings off Brora in the last twelve years, all presumably of Far Eastern rather than Icelandic origin: an immature flew NE with common scoters on 9 Oct 04, an immature or female arrived from the NE on 7 Jan 07, landing on the sea well offshore, and a more obliging female consorted with a small party of goldeneye from 17 Feb 15 to 21 Apr 15.

Long-tailed Duck
Clangula hyemalis
Common winter visitor to the Moray Firth coast, with smaller numbers in northern sea lochs and bays. Rare inland. Like the eider, this species has declined in the Moray Firth since the 1980s. *See photo right.*

Common Scoter
Melanitta nigra
The status of the small breeding population in the Flow Country appears unchanged. Passage migrant

and abundant late summer and winter visitor in the south-east, mainly off Embo/Dornoch. Maxima 700 Embo July 04 and 1200 Dornoch 10 Feb 08 Rare in the north and west.

Black Scoter
Melanitta americana
A male was seen with the wintering scoter flock of Embo/Dornoch in 1990, 91 and 93, but no recent records.

Surf Scoter
Melanitta perspicillata
There have been several sightings in winter and spring of birds with the scoter flocks off Embo/Dornoch and in Kintradwell Bay, Brora. A male off Bora on 21 Sep 14 and off Loch Fleet on 01 Oct 14 are the most recent sightings.

Velvet Scoter
Melanitta fusca
Uncommon passage migrant, mainly late summer/early autumn, and winter resident in the Moray Firth, where numbers have declined in recent years. Max. recent count 300 on 27 Mar 08.

Drake long-tailed duck in winter plumage. Virtually all long-tailed ducks seen in Britain are in winter plumage.

Goldeneye
Bucephala clangula
A small breeding population in the south appears to be stable. Abundant winter visitor throughout.

Smew
Mergus albellus
Rare visitor, with only four records since 1996: a female on Little Loch Shin, Lairg on 8 Dec 97, a male on Loch Croispol, Durness on the extraordinary date of 22 June 04, a pair at Balnakiel marsh on 22 Mar 05 and an immature flying W past Strathy Point on 13 Oct 05.

Red-breasted Merganser
Mergus serrator
Common and widespread breeding resident, with numbers augmented by immigrants in winter.

Goosander
Mergus merganser
Fairly common and widespread breeding resident.

Quail
Coturnix coturnix
Rare summer visitor, mainly May/June, with most reports from the north of the County. Single birds recorded in 4 years from 1997. 2 in 2003, 1 in 2006, 2 in 2012 & 1 in 2013.

Red-legged Partridge
Alectoris rufa
Introduced birds are now found in most parts of the County, sometimes in sizeable coveys.

Red Grouse
Lagopus lagopus
Common breeding resident. Numbers fluctuate, but generally stable. Notable cold weather count of 172 at Londornach, near Dornoch on 07 Dec 10.

Ptarmigan
Lagopus mutus
Uncommon breeding resident. A decline is expected in response to climatic change, but there is no apparent evidence for this yet.

Black Grouse
Tetrao tetrix
This uncommon species benefitted from last century's large scale afforestation and may do so again from the numerous deciduous planting schemes now being established.

Capercaillie
Tetrao urogallus
Caledonian forest regeneration in the south of the County may result in the return of this species, which still occurs just to the south in Easter Ross.

Grey Partridge
Perdix perdix
As a truly wild bird, probably now extinct in the County. Even local re-introductions do not seem to have been successful, possibly as a result of the rapid spread of red-legs.

Pheasant
Phasianus colchicus
All too common and widespread. Controls on the numbers released in the future would be beneficial, as they consume large numbers of invertebrates, especially caterpillars.

Red-throated Diver
Gavia stellata
The widespread breeding population appears to be stable. Common winter resident on coasts, particularly in the south-east.

Black-throated Diver
Gavia arctica
Breeding success has been almost doubled where floating rafts have been used, which should result in a slight increase in the breeding population. Uncommon on coasts in autumn/winter. *See photo below.*

Black-throated diver in summer plumage.

Great Northern Diver
Gavia immer
Evidence of breeding in several recent years. Summering individuals, probably mostly first year birds, complicate the picture and include 15 (one adult) off Embo in 2014. Fairly common passage migrant and winter resident, particularly on north coast, e.g. 39 flying W off Strathy Point on 3 Nov 01 and 27 in Loch Eriboll on 8 Apr 03.

White-billed Diver
Gavia adamsii
No longer the rarity it was once considered to be, with several sightings in most years. Recorded as singles from all coasts, mostly in late Sep/Oct, when birds transit Scandinavia from the Russian Arctic to winter off western Scotland. Some of these pass through the Pentland Firth, as did 2 (adult and immature) on 27 Sep 08. An adult off Brora on 4 Oct 08 circled high and flew NNW (inland), suggesting it was too far south. One off Handa island on 1 May 06 and a 1st year bird in the outer Dornoch Firth on 10 May 15 are the only spring records.

Giant Petrel sp.
Macronectes sp.
There have been two sightings of these albatross-sized petrels off Brora: an immature on 29 Oct 00 and a dark-phase adult thought to be 'Southern' M. giganteus (conceivably the same bird, stranded in the N. Hemisphere) on 25 Oct 08. Both were flying south into strong winds, well offshore.

Fulmar
Fulmarus glacialis
Breeding numbers probably peaked in the 1980s and many 'overflow' sites, particularly inland, are no longer occupied. The species is still very common in the north and west and huge passages can occur in autumn, e.g. more than 20,000 flying W off Strathy Point on 9 Sep 97 and 12,000 there on 22 Aug 03. Small numbers of 'blue-phase' birds are regular in autumn.

Fea's Petrel
Pterodroma feae
A coast-hugging bird flew NE, less than 200 yards offshore, at Brora on 29 Aug 03, having earlier been seen off Fife Ness. It later passed the northern tip of North Ronaldsay.

A second bird passed Brora and Helmsdale on 17 Sept 13.

Scopoli's and Cory's Shearwater
Calonectris diomedea/borealis
Records of this recently split pair were not assigned to race but it

is likely that both have occurred. Small numbers occur almost annually, mainly in late Aug/early Sep, with a total of 5 in 2003.

Cape Verde Shearwater
Calonectris edwardsii
The inclusion of this sub-tropical species, also formerly regarded as a race of Cory's, will understandably raise a few eyebrows as it is not on the British List, but the exceptional circumstances merit its serious consideration. In early Sep 07 a large, static high pressure system dominated the eastern Atlantic. Low pressure to the north 'squeezed' the isobars, creating a strong W/E airflow. On the western edge of the high pressure, in mid-Atlantic, the winds were strong southerly, which were responsible for the displacement of thousands of great shearwaters (normally south of the Azores by this time) back into northern waters. With the 1400+ passing Strathy Point on 8 Sep, there was only one Cory's-type. It was smaller than the greats (Cory's/Scopoli's always look longer-winged) with uniform dark brown upperparts and was slimmer and longer-tailed than either of its congeners. Critically, it was also dark-billed – another diagnostic feature of edwardsii. The observer, AV, has seen thousands of diomedea and borealis, and, more recently, an identical Cape Verde off the Azores. Incredibly, a similar bird (but almost certainly a different individual) was seen and videoed by Martin Scott off the Butt of Lewis, which logged over 7000 greats, the same day.

Great Shearwater
Puffinus gravis
Uncommon autumn migrant, mainly Aug/Sep, but recorded between 23 July and 26 Nov (when 2 flew NE together off Lothbeg Point in 05). An exceptional passage of over 1400 occurred off Strathy Point on 8 Sep 07, when more than 7000 were counted off the Butt of Lewis (see above).

Sooty Shearwater
Puffinus griseus
Fairly common, sometimes abundant, passage migrant, mainly from late Aug to Oct (latest 2 Dec 05). 494 flew NE off Brora on 28 Sep 01.

Manx Shearwater
Puffinus puffinus
Common non-breeding summer visitor and passage migrant, rare in early winter. Earliest 23 March. Largest movements Aug/Sep, e.g. 800 off Brora on 27 Aug 97 and 756 on 8 Sep 02; 2500 off Strathy Point on 22 Aug 03.

Adult gannet.

Balearic Shearwater
Puffinus mauretanicus
Uncommon but increasingly regular passage migrant, mainly Aug – Oct, recorded from all coasts. Usually in ones and twos but a total of 4 flew NE off Lothbeg Point on 10 Oct 02. Extreme dates 28 June – 1 Nov.

Macaronesian Shearwater
(Formerly Little Shearwater)
Puffinus baroli
Vagrant. There have been no sightings since 1994.

Wilson's Petrel
Oceanites oceanicus
Vagrant, but likely to occur more frequently as sea temperatures rise. Following a prolonged seabird feeding frenzy in Strathy Bay, an adult flew W past Strathy Point, close inshore, on 11 Sep 05.

Storm Petrel
Hydrobates pelagicus
Breeding summer visitor on offshore islands. Feeding and passage birds commonest in north and west, mainly July – Sep. 800 flew W in westerly gale, Strathy Point, 22 Aug 03. Early Dec singles off Brora in 98 and 05.

Leach's Petrel
Oceanodroma leucorhoa
Rare breeder (singles trapped Eilean Hoan in June and Aug 03). Uncommon passage migrant, mainly Sep/Oct. A record 51 flew W off Strathy Point on 9 Sep 97 and 9 flew W there on 8 Sep 07.

Gannet
Morus bassanus
Common non-breeder off all coasts throughout year except Jan to mid-March. Huge autumn movements off north and west coasts and c.13000 flew NE off Brora on 31 Oct 03. *See photo left.*

Cormorant
Phalacrocorax carbo
Common breeding resident, but scarce inland. A small immature at Brora on 18/19 Oct 00 was almost certainly of the eastern race *sinensis*.

Shag
Phalacrocorax aristotelis
Common breeding resident. Subject to population fluctuations, but overall fairly stable.

Bittern
Botaurus stellaris
With the breeding range having spread north to the Tay, this species might be expected to occur more frequently. One flew NW, calling, over Rhilochan, Strath Brora, on the evening of 23 May 14.

Snowy Egret
Egretta thula
A small egret seen on 29 Sep 01 by Charles and Sylvia Kataw (members of the East Sutherland Bird Group) at Colaboll, Loch Shin was not present early the next morning. When photographs became available of the Argyll snowy egret, which was found two days later, they were able to confirm it was the same bird. It is likely that the bird travelled down Sutherland's 'north-west fault' and therefore probably arrived in Britain via Iceland.

Little Egret
Egretta garzetta
Despite this species' dramatic colonisation of southern Britain, it remains a rarity in Sutherland. One at Loch Fleet on 16 Apr and 29 May 16.

Great Egret
Egretta alba
Vagrant, but likely to occur more regularly in future. One at Colaboll, Loch Shin on 28 May 03 was seen in the Dornoch Firth the next day. One at Bettyhill from 4 – 10 June 03 could have been the same individual.

Grey Heron
Ardea cinerea
Common breeding resident. *See photo below.*

Purple Heron
Ardea purpurea
Vagrant. Visitors to the RSPB Forsinard Reserve reported flushing one from a roadside ditch N of Kinbrace on 16 June 03 – a likely date, and on a well used N/S flight line. Another was reported near the north coast in late June 04. A juvenile arrived from NW and continued SE past Strathy Point on 21 Sep 03.

Grey heron in breeding plumage.

Black Stork
Ciconia nigra
Vagrant. One at Loch Fleet on 29/30 June 99 and one between Kyle of Tongue and Loch Eriboll on 12 May 17 were the second and third County records..

White Stork
Ciconia ciconia
Spring overshooters from the Continental breeding range occur occasionally. In 1998, one travelled N from Loch Shin to Durness on 22/23 Apr, returning to Loch Fleet and then Lochinver on 26th/27th. There was another at Durness on 23 Apr 06 and a party of 3 did the grand Scottish tour in 2008, calling at Loch Shin on 28/29 Apr. There have been no autumn records since the bird at Helmsdale and Brora in Oct 96.

Glossy Ibis
Plegadis falcinellus
Vagrant. No records since the only one in Dec 1962.

Spoonbill
Platalea leucorodia
Vagrant. No records since the bird at Loch Fleet in Dec/ Jan 1975/76.

Little Grebe
Tachybaptus ruficollis
Sparsely distributed breeder, wintering on low, ice-free lochs and in sheltered south-east coastal waters, like Loch Fleet.

Great Crested Grebe
Podiceps cristatus
Uncommon and irregular visitor mainly to south-east coast in winter, where Embo is a favoured locality. Recorded from Durness (late March), Loch Assynt (June) and Scourie (early Aug).

Red-necked Grebe
Podiceps grisegena
Uncommon autumn migrant and winter visitor of more regular occurrence than great crested. Almost all sightings are from the south-east coast, earliest one flying NE off Brora on 12 Sep. One Balnakeil bay, Durness, 28 Sep 98.

Slavonian Grebe
Podiceps auritus
No recent breeding attempts. Winter resident in moderate numbers in outer Dornoch Firth (max. 65 Dornoch & 37 Embo on the 13 Jan 11) and in north-west, where Loch Eriboll is favoured. Arrivals from late Sep. 19 still present in Kyle of Tongue on 8 Apr 03. *See photo overleaf.*

Slavonian grebes often spend the winter months offshore and may be seen close in during calm weather.

Black-necked Grebe
Podiceps nigricollis
Rare visitor, mainly in winter. Both recent records are from Brora, where one flew NE on 12 Oct 04 and one flew SW on 17 Dec 05.

Honey Buzzard
Pernis apivorus
Probably established as a rare but regular breeder in the south from the late 1990s.

Uncommon passage migrant in spring and autumn, earliest 26 Apr 04 (Clynelish valley, Brora. An exhausted female arrived at Lothbeg Point on 22 May 02). On the evening of 20 May 15, in classical 'overshooting' conditions, a party of 5 arrived from NW in upper Strath Brora, gained height over Dalreavoch and drifted SE, first three, then the other two. In central Sutherland, one at Achentoul on 11 July 97 and one flew S there (attacked by a resident buzzard) on 22 June 03

Autumn migrants, probably of Scandinavian origin, occur irregularly; extreme dates 5 Sep (98) and 6 Oct (08). During an unusually large influx into eastern Britain in late Sep 00, one flew SW at Loth on 23rd and there was one near The Mound the next day.

Black Kite
Milvus migrans
A marked increase in the numbers reaching Britain in recent years is certainly reflected in Sutherland, with several records since 2001 and one individual returning to Achentoul in each year from 02 to 05, when it was seen on 21 Apr but not subsequently. Earliest was one flying W through Strath Brora on 4 Apr 01.

Red Kite
Milvus milvus
The process of re-colonisation from the Black Isle reintroduction programme has been a slow one, but it is now an established breeder in the south-east. Communal roost of 21 near Loch Fleet in early 2017. *See photo below.*

White-tailed Eagle
Haliaetus albicilla
Wandering birds from the west Scotland reintroduction programme occurred with increasing frequency from the early 90s and breeding in the north-west was finally confirmed in 2015. *See photo overleaf.*

Red kites are slowly becoming established in south- east Sutherland.

Adult white-tailed eagle. This species can be expected to spread more widely within Sutherland in the coming years.

Marsh Harrier

Circus aeruginosus

With a large increase in the British breeding population and a range expansion as far north as the Tay, occurrences in Sutherland should become more frequent. There have been five records since 2002: an immature female at Loch na Claise, Stoer on 17 Aug 02; one Dornoch/Embo on 13 June 06, an adult female flying NE through the Clynelish valley, Brora on 29 Apr 07, an immature female at Dornoch Point on 15 Aug 13 and a wing-tagged immature female at Rhilochan on 22 May 16. This was from a River Tay nest.

Hen Harrier

Circus cyaneus

The small breeding population of around thirty pairs has assumed extra importance with the species' decline (sadly, human-assisted) further south. Uncommon passage migrant and winter visitor.

Pallid Harrier

Circus macrourus

Rare migrant. This species from the steppes of Asia has been occurring in Britain with increasing frequency and has been almost 'regular' on Orkney in recent years. This trend is reflected in two August sightings of juveniles: one flying S at Trantlemore in 2008 and one in the north-west in 2015.

Montagu's Harrier
Circus pygargus
The first County record was of a 1st summer female flying NE through the Clynelish valley, Brora on 28 Apr 97. The second was of a ringed juvenile found freshly dead at Invershin on 7 Oct 02. It had been rescued from a damaged nest in SE France and released from care. Its appearance in Sutherland's 'north-west fault' flight line suggests it could have been the bird seen in Iceland earlier that autumn.

Goshawk
Accipiter gentilis
Probably originating from captive stock, goshawks are well established breeders in the south and centre of the County. One at Raffin, Stoer from 7 – 9 Sep 03.

Sparrowhawk
Accipiter nisus
Common and widespread breeding resident. Two on Handa island in Sep 05 may have been migrants.

Common Buzzard
Buteo buteo
Common, widespread and increasing resident. A pair south of Dalchork had two eggs by 24 Feb in 03. Unfortunately, its success may be partly responsible for a noticeable decline in breeding waders. *See photo below.*

Buzzards remain one of the most numerous birds of prey in the area.

Rough-legged Buzzard
Buteo lagopus
Rare passage migrant and winter visitor. One flew SE through Strath Brora, 26 Jan 97; an adult male (rare in Britain) flew NE through the Clynelish valley, Brora, 27 Feb 03; one Bettyhill 17 – 23 Nov 04; another in the Clynelish valley, 18 Oct 08; one Gordonbush, 27 April 09 and 1 Forsinard 25/26 Apr 12.

Golden Eagle
Aquila chrysaetos
The resident breeding population of around 60 pairs is concentrated in the west and north-west. The last breeding pair near the east coast was displaced by an ill-sited windfarm, despite local objections.

Osprey
Pandion haliaetus
Increasing breeding summer visitor, with most pairs in the south. Extreme dates (31 March – 09 Nov). One flying SW well offshore from Brora on 17 Sep 08 may have been a migrant from Scandinavia. *See photo below.*

Ospreys frequent intertidal bays on the east coast and nearby lochs between April and August.

Kestrel
Falco tinnunculus
Fairly common breeding resident and passage migrant. After a decline in the second half of the last century, there have been encouraging signs of a recovery in the last twenty years.

Red-footed Falcon
Falco vespertinus
Vagrant. An immature female flew SE through the Clynelish valley, Brora on 5 Oct 97.

Merlin
Falco columbarius
Uncommon but widely distributed breeder and passage migrant. Winter birds are probably a mix of local ones moving to coasts and immigrants from Iceland and Scandinavia (e.g. 2 arriving from E at Lothbeg Point on 12 Oct 05). Pair near Rhilochan 'shadowed' hunting hen harrier on 11 Sep 14, watching its progress from low vantage points. Juvenile male attacked by great skua off Strathy Point, 19 Aug 05.

Hobby
Falco subbuteo
June and July records from southeast suggest it is a potential breeder. Occurs as a migrant in most years, mainly in Aug, but one arrived at Strathy Point from Hoy on 2 Oct 00, one flew SW through Clynelish valley, Brora on 30 Oct 99 and an exceptionally late bird followed the same line on 27 Nov 06.

Gyrfalcon
Falco rusticolus
Vagrant. One on Handa island on 18 Apr 97; one white phase bird near Conival (Assynt) 17 Mar 2002.

Peregrine
Falco peregrinus
Uncommon, but widely distributed breeding resident. With no prospect of 'urbanisation', the population appears stable at around 50 pairs.

Water Rail
Rallus aquaticus
Probably an occasional breeder (e.g. calling birds at Rhilochan in May 97 and 10 July 15 and a 'singing' male at dusk on 14 May 16). Balnakeil marsh is another likely site. Uncommon passage migrant; rare in winter.

Spotted Crake
Porzana porzana
Occasional breeder, probably overlooked. One calling Forsinard, June 03.

Baillon's Crake
Porzana pusilla
Vagrant, not recorded since the 19th Century.

Corncrake
Crex crex
Rare breeder and passage migrant. Conservation measures in the extreme north-west have resulted in a small increase in that population. Isolated records from elsewhere in the County include 2 calling near Golspie in Aug 03.

Moorhen
Gallinula chloropus
Uncommon and very local breeder. Now resident at Balnakeil marsh, Durness. This adaptable and opportunistic species seemed to be on the verge of local extinction at the end of the last century but has made a comeback with recent breeding successes at Rhilochan and Loch Evelix, where 10 in Oct 15.

Coot
Fulica atra
Rare non-breeding visitor. There have been few records since 1997, most recently one at Loch Evelix in Oct 13 and one at Rhilochan in May 15.

Common Crane
Grus grus
Recent breeding success in Caithness should result in many more sightings, and possible breeding attempts, in Sutherland. Birds have occurred in the Durness area several times since 1997 (March – June), with others in the south-east and centre of the County.

Sandhill Crane
Grus canadensis
Vagrant. The 2009 bird, which was first seen in the Northern Isles (South Ronaldsay), visited Brora on 29 Sep 09.

Stone Curlew
Burhinus oedicnemus
The first sighting in the County was near Brora on the seemingly unlikely date of 8 Jan 09, when 3 flew SW through the Clynelish valley. A sudden freeze in southern Scandinavia, where some birds overwinter, may have been responsible for their unseasonal arrival. One at Balnakiel from 28 – 31 May 16.

Black-winged Stilt
Himantopus himantopus
Vagrant. No records since a (disputed) sighting in 1953.

Avocet
Recurvirostra avosetta
The recent northward range expansion of this species in eastern England increases the likelihood of occurrences in Sutherland. Two in

Oystercatcher is one of the most widespread of breeding waders in the county.

the Kyle of Tongue from 13 Jan, with one remaining until at least 23 Feb 16

Oystercatcher
Haematopus ostralegus
Common breeder, passage migrant and winter resident. Hundreds were moving SW of Brora on 29 July 08. See photo above.

Grey Plover
Pluvialis squatarola
Uncommon passage migrant. Small numbers winter in the Dornoch Firth, maxima 40 Dornoch Point on 14 Oct 08 and 31 there on 9 Nov 10.

Golden Plover
Pluvialis apricaria
This protected species has suffered a serious decline, firstly from the afforestation of large tracts of moorland (which also gave predators greater access to adjacent breeding habitat) and, more recently, by the siting of wind-farms in inappropriate places. The worst example concerns a population of 42 pairs (the highest breeding density known in Britain (Dr D.A. Ratcliffe, pers. comm.)) above Gordonbush, Strath Brora, which has been decimated, as predicted by ornithologists. Climate change has added to the species' problems, as the hatching of chicks is no longer synchronised with the emergence of crane flies, which are appearing earlier.

Despite these setbacks, the species is still quite widely distributed in open areas and is a common migrant, with winter flocks on or near the coast, mainly in the south-east.

Small flocks of ringed plovers are a common sight on sandy beaches throughout Sutherland, but especially so in the south-east.

Dotterel
Charadrius morinellus
Scarce breeding summer visitor in suitable montane areas. Uncommon migrant in late summer/autumn. Records include one at Cape Wrath on 24 July 03, 2 arrivals from the east (presumably Scandinavia) at Lothbeg Point on 21 Sep 99 and an exceptionally late bird flying NE there on 23 Nov 02.

Ringed Plover
Charadrius hiaticula
The nominate race is a common breeding resident on coasts and suitable lochs and rivers inland. The darker-backed race *tundrae* is a regular passage migrant in small numbers. *See photo above.*

Semipalmated Plover
Charadrius semipalmatus
One at Balnakeil, Durness from 12–14 May 17 was the first County record.

Little Ringed Plover
Charadrius dubius
Rare passage migrant. Despite its colonisation of England in the 1950s and subsequent northward spread, not seen in the County until 2001 when one 'coasted' ENE past

Brora on 23 Apr, calling. A juvenile flew SW there on 24 Aug 05 and an adult flew NE at Lothbeg Point on 24 Oct 05.

Kentish Plover
Charadrius alexandrinus
Vagrant. The only sighting since the first at Brora in late May 1994 was one at Dornoch Point from 7 – 9 May 15.

Greater Sand Plover
Charadrius leschenaultii
Vagrant, but an increasing regular visitor to Britain due to the westerly expansion of its Asiatic breeding range. One at Dornoch Point from 16 – 24 June 11 is so far the only record.

Lapwing
Vanellus vanellus
Although still a fairly common breeder, there has been a marked decline, possibly partly as a result of increased predation of the chicks by crows and buzzards. Flocks winter on coasts, mainly in the south-east. *See photo below.*

Winter lapwing flock.

Whimbrel
Numenius phaeopus
Occasional breeder. Fairly common passage migrant, mainly Apr/May and Aug/Sep.

A very tired individual of the vagrant, dark-rumped American form *hudsonicus* was seen and photographed at Balnakeil, Durness on 20 May 12.

Curlew
Numenius arquata
Common and widespread breeder, passage migrant and winter resident.

Black-tailed Godwit
Limosa limosa
Uncommon, but increasing, passage migrant in Apr/May and Aug/Sep, mainly in the west. Parties rarely exceed 15 birds. Rare in winter. *See photo below.*

Bar-tailed Godwit
Limosa lapponica
Common passage migrant and winter resident, mainly in south-east.

Juvenile black-tailed godwit. Small numbers of black-tails occur in autumn and are mostly young birds like this one.

Summer plumage turnstone. Passage birds in May are usually in breeding plumage.

Turnstone
Arenaria interpres
Common coastal passage migrant and winter visitor, usually present in every month. *See photo above.*

Knot
Calidris canutus
Common passage migrant and abundant winter resident in south-east, max. 4200 in Feb 10.

Ruff
Philomachus pugnax
Proof of breeding was obtained in 1980 and 2015. Uncommon autumn migrant, mainly Aug/Sep. In 1998, 8 flew SW at Brora on 4 Sep and there were 10 at Durness on 21 Sep. More recently, a record 22 at Embo on 1 Sep 15.

Broad-billed Sandpiper
Limicola falcinellus
Rare passage migrant, but easily overlooked. There have been three recent sightings, all at Brora: one flew SW on 4 Sep 98 during an exceptional 'coasting' movement of Scandinavian waders, which included little stint and 8 ruffs; a juvenile flew ENE with a little stint on 19 Aug 06; an adult arrived from SW and left to ENE on 25 Apr 07.

Curlew Sandpiper
Calidris ferruginea
Uncommon autumn migrant, usually in ones and twos. An adult in breeding plumage was on Dornoch Point on 9 June 03.

Winter plumage sanderling.

Stilt Sandpiper
Micropalama himantopus
Vagrant. The bird found by the late Donnie MacDonald at Dornoch on 18 Apr 1970 remains the only record.

Temminck's Stint
Calidris temminckii
Rare breeder and passage migrant. With much suitable habitat in remote areas, the breeding population may be higher than thought. Two juveniles on the tide-wrack at Brora on 1 Aug 05 left high to the E when pushed off by the rising tide.

Sanderling
Calidris alba
Common passage migrant and winter visitor in moderate numbers, with Brora and Dornoch the most favoured localities. An exceptional count of 300 at the mouth of Loch Fleet on 17 Nov 13. *See photo above.*

Dunlin
Calidris alpina
Declining breeder; common passage migrant and winter resident, with the largest numbers in the south-east.

Purple Sandpiper
Calidris maritima
A potential breeder, as a few nest further south in the Highlands. A wintering population of long-billed birds from Greenland arrive in late Oct. Short-billed birds from Scandinavia/Russia, which winter as close as north-east Scotland, arrive earlier (e.g. 5 in from the NE at Brora on 11 July 01) but move on quite quickly.

Little Stint
Calidris minuta
Uncommon passage migrant, mainly in autumn (Aug – early Oct) although there was a moulting adult at the Kyle of Durness on 21 May 97, an adult at Dornoch on 1 July 12 and a flock of 17 on Handa island on 24 July 03. One flew NE at Lothbeg Point on 22 Nov 00.

White-rumped Sandpiper
Calidris fuscicollis
Vagrant. One at Brora from 28 – 31 Oct 12, following a major transatlantic displacement of this Nearctic species, is the only record.

Baird's Sandpiper
Calidris bairdii
Vagrant. One on Dornoch Point on 2 Nov 16 is the first County record.

Buff-breasted Sandpiper
Tryngites subruficollis
Vagrant. One at Stoerhead on 26/27 Sep 00 arrived exactly forty years after the first, at Dornoch.

Pectoral Sandpiper
Calidris melanotos
Vagrant. A juvenile at Tongue on 28 Aug 03 left to S; one on the Dornoch saltings on 17 Sep 10.

Red-necked Phalarope
Phalaropus lobatus
Former breeder, but no recent evidence, although birds have been seen in summer at a north-west site in two recent years. Two in Tarbat bay, Scourie on 25 May and one at Oldany island on 29 May 99 most likely migrants. Rare autumn migrant, but any regular offshore passage through the Pentland Firth is difficult to detect. Two juveniles flew W off Portvasgo on 1 Oct 05 and an adult female flew W past Strathy Point on 8 Sep 07.

Grey Phalarope
Phalaropus fulicarius
Uncommon but regular autumn migrant, mainly mid-July (earliest 11th) to mid Nov.

Most juveniles, which migrate later than adults, probably routinely pass through the Pentland Firth well offshore (e.g. a total of 9 off Strathy Point on 29 Oct 07). However, a high proportion of birds exiting the Moray Firth past Brora and Lothbeg Point in strong E/SE winds are adults still in breeding plumage, as were all of the record 14 which passed Lothbeg Point in ones and twos on 25 Sep 00. (Adults do not complete their moult until they reach their wintering grounds.)

Rare in winter: an adult passed Brora on 2 Dec 05, one at Golspie on 21 Dec 12 and one flew NE off Brora on 28 Dec 01.

Common Sandpiper
Actitis hypoleucos
Common breeding summer visitor, late Apr (earliest 10th in 2003) – late Sep. Post-breeding dispersal to coasts.

Greater yellowlegs which spent several weeks at Loch Fleet in 2011/12.

Greenshanks are wintering in increasing numbers at sheltered sites on the east coast.

Green Sandpiper
Tringa ochruros
Uncommon passage migrant, mainly Aug. One Clynelish valley, Brora 1 June 98.

Spotted Redshank
Tringa erythropus
Uncommon passage migrant; rare in winter. One Loch Fleet 12 Oct – 27 Nov 06 and one throughout February 14

Greater Yellowlegs
Tringa melanoleuca
Vagrant. A well-travelled individual arrived at Loch Fleet on 14 Dec 11 shortly after one was seen in Northumberland. It stayed for several weeks , then disappeared for a period during severe cold and reappeared at Dornoch on 20 Feb 12. *See photo left.*

Greenshank
Tringa nebularia
Fairly common breeder and passage migrant. Numbers of wintering birds, mainly in the south-east, have increased in recent years (max. 23 in Loch Fleet, Oct 14). High tide roost of 32, Kyle of Tongue 10 Apr 97. *See photo above.*

Wood Sandpiper
Tringa glareola
Uncommon breeding summer visitor, arriving in mid-May (earliest 10th) and very scarce passage migrant. The small number of breeding pairs fluctuates, but overall the population appears to be stable. One at Balnakeil, Durness on 22 Oct 06.

Redshank
Tringa totanus
Common, but declining, breeding resident; widespread on coasts on passage and in winter. Large numbers arrive from Iceland (e.g. at least 250 passing Strathy Point from NW on 19 Aug 05). *See photo below.*

Jack Snipe
Lymnocryptes minimus
Uncommon passage migrant and winter visitor; probably under-recorded, as much suitable habitat in remote areas. Extreme dates 18 Sep – 28 Apr.

Woodcock
Scolopax rusticola
Fairly common, but apparently declining, breeder. Some birds probably resident, but joined in winter by variable numbers of immigrants arriving from (mainly) Scandinavia in Oct/Nov.

Redshanks foraging.

Snipe are common residents.

Common Snipe
Gallinago gallinago
Common breeder and passage migrant. Many Scottish birds winter in Ireland and those in coastal sites at that season are more likely to be immigrants from Fenno-Scandinavia. *See photo above.*

Great Snipe
Gallinago media
Rare migrant, but likely to be overlooked as it prefers drier sites. One arrived from the east in a SE gale at Lothbeg Point on 11 Sep 00, settling in marram grass in the dunes. One flew south, low over Rhilochan, on 7 Oct 16, when many other migrants from the east arrived elsewhere in northern Scotland.

Pomarine Skua
Stercorarius pomarinus
Fairly common passage migrant, mainly in autumn (Aug – Nov); rare in winter. Numbers in spring are usually small but on 19 May 06 there was an exceptional overland movement (presumably through the Great Glen). Over 100 birds then crossed the mouth of the Dornoch Firth from Loch Eye, passing Embo and Brora.

Autumn flights in the Moray Firth in SE gales include a record 157 passing Lothbeg Point on 21 Oct 99. Movements off the north coast tend to be smaller, but 63 (mainly adults) flew W off Strathy Point on 27 Sep 94.

Arctic Skua
Stercorarius parasiticus
Common summer visitor with a breeding colony of around 30 pairs on Handa island and occasional breeding attempts on the northern mainland. Protracted post-breeding passage, July to Oct, no doubt includes birds from N. Isles and Scandinavia, but large movements unusual.

Long-tailed Skua
Stercoraria longicaudus
Uncommon, but regular, autumn migrant late July – mid-Oct. There has been no recent repeat of the large movements in the Moray Firth and Pentland Firth witnessed in the mid-90s (Vittery, 1997); recent max. 9 adults flying W off Strathy Point on 17 Aug 03. Rare in spring, with records from Handa island in May/June and one off Brora on 4 June 06.

Great Skua
Stercorarius skua
Common summer visitor, with a breeding colony of well over 100 pairs on Handa island. First arrivals mid-March. Autumn flights off both north and south-east coasts sometimes in excess of 100 birds. Rare inland but 2 flew NW over Rogart on 21 June 14. Single birds occasionally linger in the Moray Firth until Dec/Jan.

South Polar Skua
Stercorarius maccormicki
Vagrant. One flew NE, over the beach, at Brora on 29 Aug 14. Although obviously a 'great skua-type', it was less bulky than great, with slimmer wings and a more slender bill. The plumage was cold grey-brown, completely lacking rufous tones, with plain, greyer underparts. The pale wing flash at the base of the primaries, so obvious in great, was narrow and dull and only visible at fairly close range.

Puffin
Fratercula arctica
Common summer visitor, breeding on Handa island, the cliffs of the north-west mainland and, in the south-east, on the 'Green Table, just south of the Caithness border. Late summer/autumn coastal movements sometimes involve hundreds of birds but, exceptionally, at least 5200 flew NE past Brora on 29 July 06. Rare in winter.

Black Guillemot
Cepphus grylle
Fairly common breeding resident on all coasts. Some autumn movement noted off the Moray coast, max. 31 flying NE past Lothbeg Point on 26 Nov 05.

Razorbill carrying food.

Razorbill
Alca torda
Common breeder. Most local birds probably leave Sutherland waters and the huge autumn flights off the Moray coast (e.g. 8000+ passing Lothbeg Point on 11 Oct 06) may consist mainly of birds from the Norwegian Sea. Smaller numbers overwinter, mainly in sheltered waters. *See photo above.*

Little Auk
Alle alle
Winter visitor in variable numbers, usually from late Oct to March, but one off Brora on 16 Sep 99 and 6 off Dornoch on 28 Sep 03. Flights of hundreds can occur off the Moray coast in winter storms but there have been no recent movements to match the 10000+ which passed Brora and Lothbeg point on 5/6 Jan 95. More regular, but usually in smaller numbers, off the north coast, max. 71 flying W off Strathy Point on 29 Oct 07.

Pacific Auklets
In 'The Birds of Sutherland' (1997) AV detailed occurrences of crested auklets *Aethia cristatella* off Brora in the 1990s and postulated that, with the north-west passage remaining ice-free for much of the year and the stormier weather predicted for high latitudes (both a result of global warming), more birds from the northern Pacific would reach British waters from either the north-west or the Bering Sea to the north-east. Less regular seawatching has been done since AV's departure from Brora in 2009, so no direct comparison can now be made with the mid-90s, but there was one further sighting of a Crested Auklet after the publication of the book: an adult flying SW off Brora on 2 Aug 97.

In December 2005 little auks were passing Brora regularly in moderate numbers. On 17th, one party, close inshore, was accompanied by a very small, charcoal grey auklet with paler grey underwing coverts and a pale patch on the belly. Research of the literature came up with only one possibility: Cassin's auklet *Ptychoramphus aleuticus*! With some little auks now wintering in the Pacific, it is possible it had joined a flock as they returned to the Atlantic in spring. Submitting this as a first for Britain would be a pointless exercise, but it is included to illustrate that nothing can be ruled out as the planet's weather systems become more extreme and that the Moray Firth, with its 100 mile, north-facing 'capture zone' from Duncansby Head to Fraserburgh, is the most likely place to encounter such vagrants (see p. 98).

Guillemot
Uria aalge
Common breeder and winter resident in relatively small numbers. As with Razorbill (see above) autumn movements of thousands off the Moray coast probably originate in the Norwegian Sea.

Brünnich's Guillemot
Uria lomvia
With occasional specimens washed up on the coast, it seems clear that others must pass offshore unnoticed. Intensive seawatching (and better optics!) in recent years have established this species is a rare, but fairly regular, late autumn visitor to the outer Moray Firth. There have been ten sightings of single birds flying NE off Brora and Lothbeg Point since 2000, all between 24 Oct (when 2 passed separately in 2005) and 28 Nov with the exception of one

(with razorbills) on 12 Sep 15, an early 'northern' day when a Ross's gull also passed.

Bridled Tern
Onychoprion anaethetus
Vagrant. A juvenile was feeding off Brora on 21 Sep 08. The species has also been recorded in the Minch recently, not far outside Sutherland waters.

Little Tern
Sternula albifrons
Uncommon and declining breeding summer visitor to the Moray Firth coast (earliest 30 Apr). Until 2012 nested within the protection of an arctic tern colony at Brora (12 pairs raised 20 young there in 06). Now nesting is more sporadic. Rarely seen after the end of August.

Black Tern
Chlidonias niger
Uncommon autumn migrant on the Moray coast (mainly late Aug – mid-Sep), mostly off Dornoch, max. 5 there on 1 Sep 12. One flew W off Strathy point on 9 Sep 99.

Autumn gathering of sandwich terns.

Sandwich Tern
Sterna sandvicensis
Common and increasing breeding summer visitor, with new colonies now being established in the north and north-west. Autumn maxima 360 at Dornoch on 11 Aug 11 and 380 there on 1 Sep 12. Rare winter records probably involve wanderers from Orkney. *See photo on previous page.*

Common Tern
Sterna hirundo
Common breeding summer visitor to sheltered waters on the Moray and west coasts and some inland sites. Earliest Brora, 18 Apr 03.

Roseate Tern
Sterna dougallii
Rare migrant, with no records since 1995.

Arctic Tern
Sterna paradisaea
Common breeding summer visitor, with colonies on each coast. Late summer/autumn passage migrant through the Pentland Firth from more northerly parts of range, max. 250 flying W off Strathy Point on 9 July 03. Stragglers often remain in Sutherland waters into Nov.

Ivory Gull
Pagophila eburnea
Rare visitor from the Arctic, occurring in any season. Two recent

Adult Sabine's gull in summer plumage, showing characteristic wing markings.

records of 1st winter birds: one at Inverpolly fish farm on 7 Sep 98 and one at Farr bay, Bettyhill from 15–17 Nov 04 and then at Coldbackie, Tongue from 23 Nov until 18 Jan 05. In addition, there is a previously unpublished record of an adult flying NE past Brora with kittiwakes on 12 Aug 92.

Sabine's Gull
Xema sabini
Uncommon , but regular, autumn migrant, mainly through the Pentland Firth and usually in ones and twos. Five juveniles flew W off Strathy Point on 9 Sep 97, a record 9 (6 adults) passed Portvasgo in a severe W gale on 22 Aug 03 and 7 juveniles flew W off Strathy Point on 8 Sept 07. Rarer in the Moray Firth, but several sightings from Brora and Lothbeg Point: latest 21 Oct . *See photo left.*

Kittiwake
Rissa tridactyla
Abundant breeding resident and passage migrant; fewer in winter. Huge flights in onshore gales in autumn include a staggering 12000 per hour passing Brora on 11 Sep 03.

Bonaparte's Gull
Chroicocephalus philadelphia
Vagrant. Two recent sightings: a 1st summer bird flying NE at Brora on 2 Aug 06 and an adult at Dornoch Point from 14–17 Aug 11.

Black-headed Gull
Chroicocephalus ridibundus
Common breeding resident, but several south-eastern breeding colonies were inexplicably abandoned in the mid 90s.

Little Gull
Hydrocoloeus minutus
Uncommon passage migrant. Now also a winter resident in the south-east, possibly due to the rise in sea temperature. Numbers remain small (max. 16 at Brora on 3 Sep 04), despite the large increase in the southern North Sea, but there have been more records away from the Brora/Dornoch nucleus, e.g. at Melvich, Strathy Point and Stoerhead.

Wintering individuals at Brora, when food would have been relatively scarce, regularly employed a previously unrecorded feeding strategy. They 'adopted' a single auk (razorbill or guillemot), swimming close by until it dived and then tracking its underwater movements, which presumably brought some small edible items to the surface. (Vittery, 2001)

Ross's Gull
Rhodostethia rosea
Vagrant. Like ivory gull, can occur in Britain at any season. Three recent records: a winter plumage adult flew NE with kittiwakes at Lothbeg Point on 21 Sep 99, an adult flew NE off Brora on 12 Sep 15 and a 1st winter bird was seen off the Brora rivermouth, briefly, on 1 Jan 16.

Laughing Gull
Larus atricilla
Vagrant. The only two to have reached the County were 2nd year birds, at Dornoch in 1996 and Brora in 2006. Curiously, both were found on 13 Aug and last seen on 12 Oct!

Mediterranean Gull
Larus melanocephalus
The dramatic northward spread of this species, which is now breeding in numbers in southern Britain, is reflected in an increase in sightings since 1998, when a 2nd year bird at the Kyle of Durness on 20 Sep was considered a notable rarity. Most recent records are from the Moray coast in Aug/Sep, but also 2 at Brora in early Nov 14, one at Dornoch in Jan 15 and an overwintering adult at Brora in 2015/16.

Common Gull
Larus canus
Common breeding resident, with some large concentrations of immatures in the Moray Firth, e.g. at least 500 feeding off Brora on 13 May 00.

Mew Gull
Larus brachyrhynchus
Vagrant (not on British List). A small 1st winter gull (slightly smaller than Black-headed) feeding over the surf at Brora with common and black-headed gulls on 18 Oct 00 exhibited all the characteristics of this North American species (formerly regarded as a race of common gull): upper tail heavily barred brown with a broad brown sub-terminal bar (similar to ring-billed); medium grey mantle with darker grey on hind neck; solid, pale brown forewing from carpal joint with darker brown coverts and secondaries and narrow dark bar inside trailing edge; pale underparts more sullied than Common, with brownish wash on flanks and markings at sides of throat; small, darkish bill. The similar east Siberian race *kamschatschensis*, which is perhaps as likely, is described as being much darker, particularly on the underparts.

Ring-billed Gull
Larus delawarensis
Rare passage migrant. There were four sightings of single birds off Brora between 2002 and 2007, from late Feb to early June, all 1st year apart from an adult flying NE on 24 Mar 06.

Lesser Black-backed Gull
Larus fuscus
Breeding summer visitor and passage migrant, mainly in moderate numbers, although a total of 185 flew SW through the Clynelish valley, Brora on 5/6 Sep 97. Rare in winter, when most individuals are of the darker-backed race *intermedius*.

Slim, black-backed adults, considered to be of the nominate race ('Baltic Gull'), were seen at Brora on 2 Oct 04, 24 May 05 and 26 June 06.

Taimyr Gull
Larus (heuglini) taimyrensis
Extreme vagrant. An adult overwintered at Brora in 2011/12. The taxonomic status of this distinctive Siberian gull is uncertain. If accepted (currently under consideration by the relevant committees as a form of *heuglini*), it would be new to Britain and Europe.

Herring Gull
Larus argentatus
Abundant breeding resident. 'Loafing' and feeding flocks of over a thousand birds are often seen off the Moray coast and 1070 were counted at Laid fish farm, Loch Eriboll on 6 Feb 06. Small numbers of the darker-backed nominate race occur from Nov to Feb.

Yellow-legged Gull
Larus michahellis
This southern relative of the herring gull is occurring (or being detected) in Britain more frequently. One juvenile, Brora, 18 – 20 Oct 06 and up to 3 there between 5 Sep and mid-Nov 08. The only adult reported so far was at Bettyhill on 22 June 08.

American Herring Gull
Larus smithsonianus
Vagrant. A 1st winter bird feeding with herring and common gulls in a field at Brora on 12 Jan 08 had unmarked uniform brown underparts, a palish head with a pale, dark-tipped bill, barred rump and solid black tail. It left to the NE.

Iceland Gull
Larus glaucoides
Winter visitor in variable numbers and spring passage migrant, but recorded in every month except July. There was an influx in Jan 05 of at least 20 birds, mainly in the west, and an unprecedented arrival in Jan 12 when the County total probably exceeded 100. The largest concentration was at a fish farm west of Kylesku, where there were at least 40.

Small numbers of northbound birds are regularly seen, mainly on the Moray coast, between late April and early June. *See photo below.*

Kumlien's Gull
Larus (glaucoides) kumlieni
Uncommon winter visitor and spring passage migrant. This western form of Iceland Gull itself comes in two 'flavours' and is a likely future 'split'. First year birds of the rarer 'western' form (apparently a Kumlien's/Thayer's intergrade) occurred at Durness on 14/15 June 99 and Brora on 21 June 08. Three different birds were seen at Brora on successive

Iceland gull in winter plumage showing the all-white wing tips.

days, 19 – 21 May 02: an adult, a 2nd year and a 3rd year. Several immatures arrived with Icelands during the major influx in 2012 (see above).

Glaucous Gull
Larus hyperboreus
Winter visitor and spring passage migrant in small numbers, but recorded in every month. Several occurred during the influx of iceland gulls in 2005 (see above), including birds inland at Bonar Bridge and Lairg. Spring migrants include 3 flying NE at Lothbeg Point on 4 Apr 98 and 3 at Faraidh Head on 27 May 04. Adults are rare, but one flew SW at Brora on 19 Sep 08.

Great Black-backed Gull
Larus marinus
Common breeding resident. After a large increase during the last century, the population seems to have stabilised. Forages far inland in summer and is a threat to wader and waterfowl chicks.

Pallas's Sandgrouse
Syrrhaptes paradoxus
Sadly, there have been no sightings since the invasions of the late 19th century. We live in hope!

Rock Dove/Feral Pigeon
Columba livia
'Pure' birds still nest in the north and west.

Stock Dove
Columba oenas
Rare breeder in the south, with fledged young seen at Achavandra Muir in July 08 and possible breeding at Loch Fleet and Skibo in 2012/13. Few records of migrants: one arrived from the east at Dornoch on 6 Sep 11, 3 flew SW through the Clynelish valley, Brora on 17 Oct 05 and 12 flew S over Rhilochan on 9 Oct 16. In winter: 8 at Kirkton, Golspie on 8 Jan 03 and 3 at Ospisdale on 8 Dec 07.

Woodpigeon
Columba palumbus
Common breeding resident wherever suitable habitat exists. Large feeding flights in south-east in late autumn/winter mainly of local origin, although some immigration from Scandinavia possible.

Collared Dove
Streptopelia decaocto
Now a common breeding resident in towns and villages throughout, following its' first arrival in the County in 1964. *See photo overleaf.*

Collared doves.

Turtle Dove
Streptopelia turtur
Rare migrant, and likely to become even rarer due to the collapse of the British and north European breeding populations. Only three records of single birds in May/June since 2002 and one Aug sighting at Brora in 07.

Cuckoo
Cuculus canorus
Common summer visitor, late April – July. In contrast to this species' decline in southern Britain, the north Highland population has increased in recent years. No less than 40 were counted on a hundred mile road circuit of the west and north in May 15.

Barn Owl
Tyto alba
Despite occasional setbacks in hard winters, this breeding resident has increased and spread north over the last twenty years. Any nesting boxes or platforms left in abandoned buildings are therefore likely to be used.

Snowy Owl
Nyctea scandiaca
Rare visitor, most likely in winter. One near Tarbat, Scourie from 19 – 29 June 06 and one near Dalnessie, Lairg on 18 Feb 16 are the only recent sightings.

Tawny Owl
Strix aluco
Fairly common breeding resident in lowland areas, mainly in south and south-east of the County.

Long-eared Owl
Asio otus
The most widespread breeding owl, occupying plantations, native woodland, scrub and small copses in largely tree-less areas as far as the north-western extremity of the County. Sites at higher altitudes in the hinterland seem to be abandoned in winter. Some migrants from Scandinavia pass through in late autumn (e.g. 1 flying SW over the sea off Brora on 17 Oct 05). *See photo right.*

Short-eared Owl
Asio flammeus
Fairly widespread breeding summer visitor, favouring open areas at mid-altitude. Dispersal to coastal areas evident in Aug and migrants from Scandinavia responsible for some coastal Sep/Oct sightings. Uncommon in winter.

Nightjar
Caprimulgus europaeus
Formerly bred in the south-east. One churring at Gordonbush, Loch Brora in May 15 raises hopes that global warming may encourage its return.

Common Swift
Apus apus
Breeding summer visitor, in only moderate numbers, in the south, and declining in some villages. Uncommon migrant further north. 12 came in from Hoy at Strathy Point on 4 Sep 97.

[An unidentified swift at Helmsdale on 31 Oct 04 coincided with an influx of pallid swifts *Apus pallidus* into Britain.]

Long-eared owls are much overlooked but are resident in Sutherland.

Wrynecks are scarce passage migrants.

Hoopoe
Upupa epops
Uncommon migrant, reported in most years in spring (Apr/May) and occasionally in autumn (Sept/early Oct, but one at Talmine on 5 Nov 03). The majority of sightings are in the west and north.

Bee-eater
Merops apiaster
Rare summer visitor which might be expected more frequently as breeding events occur further north in Britain. In 2002, 3 were seen north-west of Lairg, on Handa island on 13/14 June and at Strathy on 18 July. One at Forsinard in May 07.

Roller
Coracias garrulus
Vagrant. No sightings since the only County record in June 1979.

Kingfisher
Alcedo atthis
Now an established breeder in the south and likely to spread further north in future. Rare autumn migrant from Scandinavia. One overwintered on the River Evelix from Oct 15 to Feb 16.

Wryneck
Jynx torquilla
Rare summer visitor (a potential breeder) and passage migrant, with birds in the Clynelish valley, Brora

on 16 June 01 and in the north-west on 19 May 04. One in the Clynelish valley on 7 Sep 08. *See photo left.*

Green Woodpecker
Picus viridis
A potential colonist, with unconfirmed recent reports from the south.

Great Spotted Woodpecker
Dendrocopos major
Now a common and increasingly widespread breeding resident, as plantations give it access to new areas. Uncommon autumn migrant from Scandinavia, sometimes arriving with 'winter' thrushes.

Golden Oriole
Oriolus oriolus
Vagrant. A male was reported from Kildonan in late June 1997 and a female was photographed at Tongue in late May 03.

Red-backed Shrike
Lanius collurio
Now a regular summer visitor in small numbers, with breeding suspected in several years since 1997. Most records are in June and probably refer mainly to wandering, unpaired birds, as they come from all parts of the County. A juvenile was found dead at Durness on 26 Aug 01.

Lesser Grey Shrike
Lanius minor
Vagrant. An adult at Rhicarn, Lochinver on 5/6 Oct 98 is the only record.

Great Grey Shrike
Lanius excubitor
Uncommon (less than annual) passage migrant and winter visitor. Since 1997 there have been only six records between 7 Nov and 21 Mar, when one at Dalchork, Loch Shin in 2003. Most recently, one in the Clynelish valley on 7 Nov and one at Kildonan on 20 Nov 15.

Woodchat Shrike
Lanius senator
Rare passage migrant. One at Raffin, Stoer from 20–27 July 05 is the only record since the two birds in June and Sep 96.

Magpie
Pica pica
There is a small resident population in the Melvich area on the north coast. Otherwise an uncommon visitor in the south and south-east, with some long-stayers near Brora and Rhilochan.

Until recently jays were largely passage birds from Scandinavia but are now breeding in the County.

Jay
Garrulus glandarius
Formerly a very rare visitor during irruptive dispersals of Continental birds of the nominate race. The range of the British race *rufitergum* has now extended northwards. After sightings of singles in Assynt and near Melvich in 98 and 04, the south-east was colonised by resident birds in 2007 and breeding proven in 2015. *See photo above.*

Nutcracker
Nucifraga caryocatactes
Vagrant. Has occurred only as an irruptive migrant from (probably) Siberia, but no recent records.

Jackdaw
Corvus monedula
Abundant breeding resident. A large roost in Ben Bhraggie Woods, Golspie is shared with rooks.

Several individuals of the nominate Continental race, showing prominent silvery half-collars, have interbred with local birds in at least four locations in south-east Sutherland since the first (at Brora) in the late 1990s. The offspring have 'shadow' half-collars.

Rook
Corvus frugilegus
Common and widespread resident, except in tree-less areas. Large

roost in Ben Bhraggie woods, Golspie. This includes birds from Caithness, many of which take a short-cut across the sea in the evening.

Carrion/Hooded Crow
Corvus corone/cornix
The continued separation of these two forms into full species seems inexplicable in Sutherland, where there is complete intergradation between them. Pure 'Hoodies' are now restricted to the remoter hinterland areas, with Carrions gradually achieving dominance elsewhere, more by interbreeding than displacement.

Raven
Corvus corax
Fairly common and widespread breeding resident, with winter flocks sometimes exceeding 100 birds.

Goldcrest
Regulus regulus
Common breeding resident. One of the few species to benefit from the spread of plantations. Autumn influxes, such as that reported from the north-west on 11 Oct 03, almost certainly migrants from Scandinavia.

Firecrest
Regulus ignicapillus
Rare migrant, which overwintered in 1972, 1988 and 1990. More recently, 2 at Dalcharn, Tongue from 16–20 Dec 10. One 'filtered' SW through the Clynelish valley, Brora on 14 Oct 01, coincident with an arrival of blackcaps from east/central Europe.

Blue Tit
Cyanistes caerulescens
Common breeding resident in the southern half of the County. Sporadic visitor to the north, mainly in autumn/winter, but 11 at Loch Duartmore, Kylesku, on 6 Feb 05.

Great Tit
Parus major
Similar distribution to blue tit (see above) but even rarer in the north. One on Handa island from 21 July into the autumn in 03, when also singles at Durness in Oct/Nov.

Crested Tit
Lophonanes cristatus
Breeds in extreme south (Ardgay), and should benefit from current attempt to restore and link Caledonian Forest remnants. Occasional sightings elsewhere in south mainly in late summer, suggesting post-breeding dispersal of juveniles.

Coal Tit
Periparus ater
Another beneficiary of plantations but still scarce in the far north. One on Handa island on 20 July 03 and one at Portskerra on 24 Sep 05. Some autumn movement noted on the south-east coastal flight line, e.g. 15 (in three parties) flying SW over Braambury Hill, Brora on 16 Oct 01.

Skylark
Alauda arvensis
Common, but possibly declining, breeder which may be adversely affected by extensive 'native' plantings now being subsidised in breeding areas. Birds occupy inland territories in early spring. Some linger in upland areas well into the autumn before moving south, or to coasts, where small numbers overwinter. Hard weather movements in winter probably involve mainly birds from the large agricultural areas in Caithness, where it is much commoner.

Shore Lark
Eremophila alpestris
Possibly rare, sporadic breeder in mountainous north-west, but a pair at Ben Armine in late June 97. Equally rare passage migrant: one on Handa island on 27 June 98; 2 at the Kyle of Tongue on 12 May 03.

Swallow brood.

Short-toed Lark
Calandrella brachydactyla
Vagrant, but autumn transients from Northern Isles (where regular) probably overlooked. One flew south through the Clynelish valley, Brora, calling, on 14 Sep 03.

Sand Martin
Riparia riparia
Common breeding summer visitor, usually arriving in late March (earliest 24th). Feeding birds mass over water in cold spring weather, e.g. 450 at Loch Brora on 21 Apr 04.

Swallow
Hirundo rustica
Common breeding summer visitor. Extreme dates 19 Mar – 11 Nov. *See photo left.*

House Martin
Delichon urbicum
Fairly common breeding summer visitor, usually arriving later than the other hirundines but 2 at Golspie on 4 Apr 04. Latest 3 at Brora on 8 Oct 13.

Red-rumped Swallow
Cecropis daurica
Rare passage migrant. One at Bettyhill on 15 Mar 98 and one flying SW through the Clynelish valley, Brora on 2 June 03 are the only records.

Long-tailed Tit
Aegithalos caudatus
Common breeding resident, except in the tree-less hinterland and the north. A bird of the northern (nominate) race was with resident birds near Astle in Jan 06.

Pallas's Warbler
Phylloscopus proregulus
Rare autumn migrant. In common with several other Asiatic breeders, this species is occurring in Britain in ever-increasing numbers. Two recent records: one at West Clyne, Brora on 25 Oct 02 and one at Talmine on 11 Oct 03.

Yellow-browed Warbler
Phylloscopus inornatus
Uncommon but increasing autumn migrant. One in the Borralie plantation, Durness, 22 Oct 01; one at Lochinver, 6 Oct 06; one at Melvich, 13/14 Oct 06; 2 trapped at Dalchork, 20 and 26 Sep 15; no less than 8 in Durness area, 24 Sep 15 and one at Embo on 27/28 Sep 15.

Wood Warbler
Phylloscopus sibilatrix
Uncommon and localised breeding summer visitor in the south; rare elsewhere. One singing Kyle of Tongue 22 June 02; one trapped at Melvich 3 Oct 04.

Willow warbler is the commonest warbler in the north.

Chiffchaff
Phylloscopus collybita
Increasing breeding summer visitor in the south. Uncommon passage migrant elsewhere. Birds of the Fenno-Scandinavian race *abietinus* likely in autumn, although one in the Clynelish valley, Brora on 12 Nov 08 was nominate *collybita*.

Siberian Chiffchaff
Phylloscopus tristis
Vagrant. A bird showing the characteristics of the form '*fulvescens*' was on Dornoch Point on 12/13 Nov 08.

Willow Warbler
Phylloscopus trochilus
Abundant breeding summer visitor, early Apr to Sep. One reported from Lochinver on 30 Nov 03, but confusion with '*fulvescens*' Siberian chiffchaff possible. *See photo above.*

Blackcap
Sylvia atricapilla
Increasing breeding summer visitor. Winter migrants, arriving mid-Oct, migrate to west and north-west Europe from east/central Europe. *See photo right.*

Garden Warbler
Sylvia borin
Increasing breeding summer visitor, with most marked range extension through the 'native' deciduous woodlands in the west as far north as Scourie. Extreme dates 9 May – 29 Sep.

Barred Warbler
Sylvia nisoria
Uncommon autumn migrant. Three recent records (all juveniles): one trapped Coul Links, 1 Sep 02; one Balnakeil, Durness, 23 Sep 02; one Embo 28 Sep 12.

Lesser Whitethroat
Sylvia curruca
Pair feeding 2 or 3 newly-fledged juveniles on Clynelish Muir, Brora on 17 July 98 was then the most northerly breeding record in Britain (but has since bred in Caithness).

Increasing spring passage migrant, earliest Achmelvich 18 Apr 03. Singing males heard regularly in south-east in May/June, including 3 in Clynelish valley, Brora on 5/6 June 08. Uncommon autumn migrant: one trapped Melvich 20 Oct 02; another there, 11 Oct 09; one Embo 26 Sep 12.

Female blackcap.

Subalpine Warbler
Sylvia cantillans
Vagrant. One was reported singing at Loth on 25 June 97.

Common Whitethroat
Sylvia communis
This species does not seem to have fully recovered from the population crash caused by the Sahel drought in the late 1960s/70s. Still a fairly common, but localised, breeding summer visitor, May to Sep, in the south. One Helmsdale, 4 Oct 03.

Grasshopper Warbler
Locustella naevia
Uncommon breeding summer visitor in variable numbers, mainly in the south. Becoming more regular in response to global warming.

Booted Warbler
Iduna caligata
Rare passage migrant. The westerly extension of the breeding range accounts for the increase in sightings of this Asiatic warbler in Britain. One 'filtered' SW through the Clynelish valley, Brora on 28 Sep 00

Waxwings occur in small numbers most autumns but influxes are only periodic.

and one behaved in an identical manner, tree-hopping to the SE on this occasion, at Rhilochan, Rogart on 6 June 14.

Icterine Warbler
Hippolais icterina
Probably now a rare breeder, as several singing males have remained in suitable habitat in May/June and one was seen carrying food and 'alarming' at Amat on 28 June 09. Three recent records of migrants from the Clynelish valley, Brora: one on 6 May 02, one (singing) on 30/31 May 08 and one juvenile on 9 Sep 08.

Melodious Warbler
Hippolais polyglotta
Vagrant, but likely to be overlooked. One juvenile with a mixed bird party in a small orchard in the Clynelish valley, Brora on 13 Aug 00.

Sedge Warbler
Acrocephalus schoenobaenus
Common and fairly widespread breeder in suitable habitats, late Apr – Sep.

Reed Warbler
Acrocephalus scirpaceus
Surprisingly, this species (a regular autumn migrant on the Northern Isles) was not recorded in the County until 1997, when there was one at Brora on 11 Aug. Since then: one Balnakeil, Durness, 4 Oct 98; one singing at Lairg on 9 June 03; one at Strathan, Tongue from 7–9 Sep 03; one Dornoch Point, 25 Aug 14.

Waxwing
Bombycilla garrulus
Eruptive autumn visitor, usually in late Oct/Nov, with birds usually moving on quite quickly. Largest recent invasions in 2004 and 2008, e.g. 300 Golspie and 100 Lochinver on 28 Oct 04 and 150 Brora on 17/18 Nov 08. Uncommon in mid-winter and rare in spring. One singing in the north-west on 21 May 03. *See photo left.*

Nuthatch
Sitta europaea
Vagrant, but a potential future colonist as its range extends northwards. One visiting a feeder at Overscaig, Loch Shin on 5 May 08 is so far the only sighting.

Treecreeper
Certhia familiaris
Widespread breeding resident apart from tree-less parts of hinterland and extreme north-west. One (on cliff!) on Handa island, 28 July 03.

Wren
Troglodytes troglodytes
Common and widespread breeding resident. *See photo right.*

Starling
Sturnus vulgaris
Common breeding resident. Large flocks of migrants in easterly winds in autumn probably mainly of Scandinavian and Russian origin.

Rose-coloured Starling
Pastor roseus
Eruptive visitor from south-east Europe and Asia. The distinctive adults are likely to be reported by non-birders and there are records in most years, mostly in June/July (three in 2002). There was a juvenile at Strathy Point on 23 Aug 98. Another arrived with redwings in the Clynelish valley, Brora on 21 Oct 98 and was later relocated at a feeder in nearby Badnallan, where it stayed until 30th, feeding mainly on cheese.

Daurian Starling
Agropsar sturninus
Extreme vagrant. A male showing signs of immaturity (including a heavily mottled rump), was found at Balnakeil, Durness with immigrant Starlings on 25 Sep 98. It was seen again at Oldshoremore about thirty miles to the south, on 28th. Despite its arrival with 'Russian' starlings in strong easterly winds, which brought other eastern vagrants to northern Scotland), it was placed only in Category 'D' of the British List.

Dipper
Cinclus cinclus
Breeding resident. Although it seems to have disappeared from some river systems, it is still fairly widely distributed.

White's Thrush
Zoothera dauma
Vagrant. Since the bird at East Clyne, Brora in late Sep 91 the only record is of one, amazingly 'camera-trapped', in Migdale Woods on 31 Jan 13.

Ring ouzel
Turdus torquatus
This uncommon and now very localised breeding summer visitor has suffered a major decline in the last thirty years, in common with other regions in Scotland. Competition from the increasing mistle thrush population may be a factor. Autumn migrants (latest 6 Nov) include individuals from Scandinavia, arriving with flocks of 'winter thrushes'.

Wren.

Blackbird
Turdus merula
Common breeding resident. Large influxes of migrants from Scandinavia can occur in late autumn (e.g. 22 Oct 01, with hundreds in open moorland and mountain passes).

Black-throated Thrush
Turdus atrogularis
Vagrant. One trapped at Melvich on 2 Oct 10.

Fieldfare
Turdus pilaris
Rare and sporadic breeder. Common autumn migrant, particularly on the south-east coastal flight line, max. 8000 flying SW on 22 Oct 98. Smaller numbers in west, but hundreds at Achmelvich in mid-Oct 03. Flocks overwinter in milder years. *See photo overleaf.*

Song Thrush
Turdus philomelus
Common summer resident with birds returning from winter quarters further south (e.g. Ireland) to breeding territories from late Feb. Migrants from Scandinavia pass through in autumn and there are small numbers of winter residents in some coastal villages.

Fieldfare feeding on rowan berries.

Redwing

Turdus iliacus

Abundant autumn migrant, although there have been no recent movements to match that on 30 Oct 95, when at least 125,000 flew SW through the Clynelish valley during the day, and many more that night. More recent maxima there of over 10000 on 15 Oct 01 and 6000 in one hour on 17 Oct 05. Variable numbers overwinter and can include birds of the darker Icelandic race, which also pass through in spring.

Mistle Thrush

Turdus viscivorus

Common and increasing breeding resident. Sizeable flocks can occur in autumn, e.g. 73 in the Clynelish valley, Brora on 25 Sep 03.

Spotted Flycatcher

Muscicapa striata

Still a fairly common breeding summer visitor. Usually arrives in third week of May (but 7th in 2008) and lingering juveniles regularly seen well into Sep. *See photo right.*

Robin

Erithacus rubecula

Common and widespread breeding resident.

Bluethroat

Luscinia svecica

Potential rare breeder, having been seen in suitable habitat in the past. Rare spring migrant, most recently one at Strathy Point on 29 May 99. No autumn records.

Red-breasted Flycatcher
Ficedula parva
Rare autumn migrant, with no records since 1988.

Pied Flycatcher
Ficedula hypoleuca
Uncommon passage migrant, mainly in May and Sep. A male near Loch Assynt on 6 Apr 15 had probably overshot its breeding grounds in a strong southerly wind.

Black Redstart
Phoenicurus ochruros
Uncommon spring passage migrant. Rare in autumn and winter: one Golspie, 14 Nov 02; one Brora 29 Oct 06; one Brora from 30 Dec 16 to 20 Mar 17.

Common Redstart
Phoenicurus phoenicurus
Fairly common breeding summer visitor (Apr – Sep), mainly in southern and western woodlands. Occurs more widely as a migrant in Sep.

Whinchat
Saxicola rubetra
Fairly widespread but declining summer visitor, arriving in late Apr/early May. May benefit from current deciduous planting schemes. Uncommon autumn migrant, latest Brora 22 Sep 97.

Spotted flycatcher.

Stonechat
Saxicola torquata
Common and widespread breeding resident, but numbers fluctuate greatly in response to the severity of the winters. Hard weather concentrations in coastal areas, e.g. 22 on Brora beach on 11 Feb 01.

A male of the distinctive Continental race *rubicola* bred with a (presumed) female *hibernans* at Loch Buie in 2000, raising two broods. The same bird, or a male with *rubicola* characteristics from one of the earlier broods, was also present in 2002. Two males showing typical *rubicola* features, including large white collars and wing patches, and white rumps, all contrasting sharply with the blackish upperparts, were paired with (presumed) *hibernans* females in the same area in May 2017, suggesting that *rubicola* genes have persisted with minimal dilution in this population for at least two or three generations.

Northern Wheatear
Oenanthe oenanthe
Common and widespread breeding resident, arriving from late March. Passage migrants include some 'Greenlands' *O. o. leucorhoa*. A leucistic example of this race on Handa island on 9 May 02 could easily have been mistaken for an isabelline *O. isabellina*! Stragglers are seen well into October and there was one at Dornoch on 17 Nov 15. See photo left.

Desert Wheatear
Oenanthe deserti
Vagrant. One at Durness from 24–28 Dec 08 is the only record.

Wheatears are amongst the first spring migrants to appear.

Dunnock
Prunella modularis
Common and widespread breeding resident. In moorland areas, takes advantage of any isolated clumps of gorse.

House Sparrow
Passer domesticus
Common and apparently increasing breeding resident in lowland areas, e.g. a sudden colonisation of inland crofts near Brora in 2001. 200 estimated at Embo in late July 03.

Unidentified red seed-eaters reported from Kinlochbervie in Sep 04 proved to be erythristic house sparrows which had been feeding on salmon food stored under nets on the quayside (Vittery, 2005).

Tree Sparrow
Passer montanus
After becoming almost extinct in the County, a welcome recolonisation, with reports since 2002 in all seasons of birds in the south-east, Lairg area and Assynt. In the north, 2 at Portskerra on 12 May 05. *See photo below.*

Tree sparrows are surprisingly scarce in Sutherland, despite good populations in parts of Ross-shire.

Yellow Wagtail
Motacilla flava
Uncommon passage migrant, mainly in May. Most birds occurring in the County seem to be nominate *flava* but there was a male *thunbergi* at Clachtoll on 24 May 97 and another at Balnakeil, Durness in mid May 01. Single *flavissima* males were at Brora on 10 May 99 and Balnakeil, Durness on 29 May 16. One there on 16 July 16 may have been the same bird. In autumn, one at Balnakeil on 1 Oct 01.

Grey Wagtail
Motacilla cinerea
Fairly common summer breeder, but rare in winter.

Pied/White Wagtail
Motacilla alba
Common summer breeder; uncommon in winter, but 50 at Dornoch on 23 Dec 03. Nominate *alba* is a fairly common migrant, more numerous in autumn, although 37 at Dornoch Point on 2 May 13.

Richard's Pipit
Anthus richardi
Rare passage migrant. There have been three recent sightings: one in the Clynelish valley, Brora on 22 March 00, one on Strathy Point on 12 Sep 07 and one on Dornoch Point on 5 Feb 13.

Tawny Pipit
Anthus campestris
Vagrant. One was seen on bare ground, where sand and straw had been stored, at Rhilochan, Rogart on 12 July 14. When disturbed it left to the east and would have reached the coast via Dunrobin Glen.

Olive-backed Pipit
Anthus hodgsoni
Vagrant, but being recorded more regularly (some multiples) in the Northern Isles. One arrived from NE in the Clynelish valley, Brora on 29 Oct 98, drank at a small pool in birches, and continued to SW, calling.

Tree Pipit
Anthus trivialis
Uncommon summer breeder in south. A few migrants occur in spring easterlies, but scarcer in autumn.

Meadow Pipit
Anthus pratensis
Abundant and widespread breeder and common passage migrant. Uncommon in winter, when mainly on coasts.

A very unusual individual in dunes near the tip of Dornoch Point in Jan 12 was much paler and greyer than 'normal' winter birds, flirted its strikingly white outer tail feathers like a reed bunting, remained completely silent (even when flushed) and, over several days, never flew more than two feet off the ground. Given the clinal variation in nominate *pratensis*, with paleness and greyness increasing to the east, it was probably from the eastern extremity of the species' range, in the vicinity of the Ural Mountains.

Red-throated Pipit
Anthus cervinus
Rare migrant, easily overlooked. One flew W through Strath Brora, calling, on 9 May 08 during a period of high pressure and easterly winds. One was then seen on Handa island (conceivably the same bird) on 12 May.

Rock Pipit
Anthus petrosus
Common breeding coastal resident. Some local movement indicated by an exodus of 11 'excited' birds (4, 3, 2, 2) leaving Strathy Point high towards Hoy on 13 Oct 05.

The paler Scandinavian race *littoralis* is an uncommon passage migrant and winter visitor, and can occur in a variety of habitats including moorland, e.g. one on a frozen sheep track on Socach Hill, Brora on 27 Dec 05 and one near Astle on 18 Dec 06. There were 3 together on Dornoch Point on 24 Oct 07, where singles were also seen from 11 Mar – 3 Apr 14 and on 3 Jan 15.

Water Pipit
Anthus spinoletta
Rare visitor, possibly overlooked. One on Dornoch Point from 24 Jan – 13 Feb 15 and one at Sputieburn on 21 Oct 16 are only the second and third County records.

Brambling
Fringilla montifringilla
Rare breeder. In 1997 a pair seen feeding 2 young in 'west Sutherland' on 30 May and there was a female close to a nest with begging young near Inverkirkaig on 4 Aug. A possible brambling/chaffinch hybrid at Saval, Lairg in June 04.

Autumn migrant (mainly late Oct, but 2 Achmelvich on 28 Sep 03) and winter visitor in variable numbers. Large influx in winter of 03/04: 600 at Forsinain in early Jan and a record 1700 at Trentham, Dornoch on 27th. *See photo overleaf.*

Chaffinch
Fringilla coelebs
Common breeding resident. Large flocks in winter (often mainly males) include many immigrants from north-east Europe, max. 2000 near Lairg on 8 Nov 02.

Bramblings turning into summer plumage are relatively common but breeding is sporadic

Hawfinch
Coccothraustes coccothraustes
Uncommon and unpredictable migrant, with singles reported, mainly in coastal areas, in most years. Most occur in May; the few autumn records are in Oct/Nov.

Common Rosefinch
Erythrina erythrina
Has bred (1990), but the expected colonisation has not materialised, although there were singing males at Golspie in July 97 and on Eilean Hoan on 24 Aug 03. An adult female was trapped at Melvich on 31 May 03.

Rare passage migrant, but probably overlooked: 2 juveniles 'coasted' NE at Helmsdale on 7 Sep 03.

Pine Grosbeak
Pinicola enucleator
Coincident with a large irruption of two-barred crossbills into the Northern Isles, a male flew steadily west (deeply undulating flight) through Strath Carnaig on 30 July 13.

Bullfinch
Pyrrhula pyrrhula
Fairly common breeding resident, mainly in the south and west. Immigrants of the larger, nominate Continental race can occur in small numbers from mid-Oct, as in winter of 04/05 when 15+ reported, with stragglers remaining until late March and one at Scourie as late as 1 June.

Trumpeter Finch
Bucanetes githagineus
Vagrant. The bird seen on Handa island in June 1971 remains the only record.

Greenfinch
Chloris chloris
Common breeding resident. Flocks assemble in winter, often with other seed-eaters, maxima 100 Skelbo, Loch Fleet on 13 Feb 12 and, in the north-west, 150 at Balnakeil, Durness on 10 Nov 03.

Linnet
Linaria cannabina
Common and increasing breeding resident. Large winter concentrations, mainly in the south-east, maxima, 800 in the Clynelish valley, Brora on 30 Dec 03 and 900+ at Skelbo, Loch Fleet in Jan 16.

Twite
Linaria flavirostris
Sparsely distributed breeding resident from coasts to uplands. A characteristic bird of Sutherland's 'hostile' hinterland.

Even up to the publication of the BTO Migration Atlas in 2002. Scottish birds were thought to be mainly sedentary but recent ringing and trapping programmes in Orkney, Sutherland, Grampian and Wester Ross have revealed a considerable degree of movement. Substantial numbers of mainland birds move north to Orkney after the breeding season, whilst others have been re-trapped around the Moray Firth, in Aberdeenshire and the Montrose Basin. In the west, a few moved south as far as Lancashire and North Wales. Colour-ringed birds were sighted in winter in East Lothian, Northumberland and at Loch Ryan. Breeding birds have shown a high degree of site fidelity. (A.R. Mainwood)

55 'coasting' SW at Brora on 29 Oct 08 is the largest visible movement reported.

Winter concentrations include 700 at Ardgay on 6 Dec 97 and 400 at Balnakeil, Durness on 24 Feb 04.

Redpolls ssp.
Although the BOU announced in 2017 that lesser redpoll *Acanthis cabaret* would no longer be regarded as a separate species, the three 'forms' of *A. flammea* occurring in Sutherland (two of which breed or have bred) are so distinctive in the field and have little or no interconnectedness, they are treated separately below.

Lesser Redpoll
Acanthis (flammea) cabaret
Common and widespread breeding resident. Flocks in excess of 100 birds occur in winter, max. 400 in Strath Brora in Dec 15. *See photo below.*

Greenland Redpoll
Acanthis (flammea) rostrata
A likely future 'split', this distinctive, dark and large-billed form is an uncommon autumn migrant: one at Durness on 28 Sep 98 and one there on 26 Sep 99; one at Loch Brora on 7 Oct 98 and one in the Clynelish valley, Brora from 29 Oct – 14 Nov 99.

Mealy (Common) Redpoll
Acanthis flammea
Has bred in several recent years at one south-east site since the major irruption of this species from the north-east core of its breeding range in the winter of 99/00, when millions reached Continental Europe and hundreds wintered in Sutherland. At least 11 pairs bred in a loose colony in 2000, raising around 40 young in two broods.

Lesser redpoll.

Site fidelity (well documented in *flammea*) would account for repeat breeding in subsequent years up to at least 2005. Four juveniles were trapped at a nearby site in Aug 03, a pair (male song flighting) was seen in the Clynelish valley Brora on 27 June 06 and there was a male at Sciberscross on 6 June 14.

In non-irruption years, an uncommon passage migrant (mainly late autumn) and winter visitor. Mid-September birds (parties of up to 15) in the Clynelish valley in 00 and 08 were most probably a result of dispersal from local breeding sites.

Arctic Redpoll
Acanthis hornemanni
Rare winter visitor, possibly overlooked. Up to 3 (one adult) in the Clynelish valley, Brora from 24 Jan – 3 Feb 00.

Two-barred Crossbill
Loxia leucoptera
Rare passage migrant and winter visitor, although has recently bred just to the south of the County in Easter Ross.

After an influx in the Northern Isles, a juvenile was seen with redwings, feeding on rowan berries, in the Clynelish valley on 22 Oct 08. A female/1st winter was in larches at Dalreavoch, Strath Brora, on 4 Jan 16.

Common Crossbill
Loxia curvirosta
Breeding resident in fluctuating numbers. Birds of the nominate Continental race move through in irruption years, mainly June – Aug. Uncommon migrant outside breeding areas: e.g. one with goldfinches behind Brora beach on 13 Sep 01 and one on Handa island on 30 Mar 02.

Scottish Crossbill
Loxia scotica
Uncommon breeding resident in south, but wandering parties, usually of less than ten birds, frequently encountered further north, as far as Strathnaver, in search of ripe cones.

[Although a form of parrot crossbill *Loxia pytyopsittacus* is now known to be resident in ancient Caledonian Forest to the south of the County, it is unlikely to occur in Sutherland. However, Continental *pytyopsittacus* (arguably a different species) is irruptive, not dependent on natural Scots Pine and may have been overlooked.]

Goldfinch
Carduelis carduelis
There has been a dramatic increase in the breeding population of this species in the last two decades, as well as range extensions to the north and north-west and winter residency, even in some upland areas – all probably attributable to climate change.

Siskin
Spinus spinus
Common breeding resident, with large post-breeding and winter concentrations, mainly in the southern half of the County.

Snow Bunting
Plectrophenax nivalis
Scarce breeder on the higher mountains in central and western areas.

Fairly common winter visitor arriving in late Sep/Oct (one at Stoerhead 19 Sep 03), with largest flock usually on Dornoch Point (e.g. 120 on 6 Jan 00) but 150 in the Clynelish valley, Brora on 2 Mar 04. Hard weather arrivals in mid-winter (e.g. 75 flying SW at Brora on 21 Feb 99) may originate from no further than Caithness. Uncommon in spring; a male on Handa island from 22 May – 2 June 06. *See photo right.*

Lapland Bunting
Calcarius lapponicus
Uncommon passage migrant (late Sep/Oct and Apr/May) but recorded less than annually. Three flying SW at Brora on 16 Oct 01 was the largest number recorded until an exceptional influx into Scotland in 2010, when 71 near Kinlochbervie on 2 Sep and 7 at Dornoch on 17 Sep. Three flew SW at Lothbeg Point on 23 Oct 11.

Black-headed Bunting
Emberiza melanocephala
Rare migrant, occurring more frequently in Britain, but only two sightings since 1995: a male at Kinlochbervie on 29 Aug 97 and a male in the Clynelish valley, Brora on 15 June 99, which left to the south.

Red-headed Bunting
Emberiza bruniceps
Birds occurring in Britain are generally regarded as escapes from captivity, although genuine vagrancy is possible. There was a male near Brora from 11 – 14 June 98 and one (possibly the same) in the north-west at Scouriemore on 7 July 98.

Corn Bunting
Emberiza calandra
Extinct as a breeding bird in the County since the mid-1970s. No recent records.

Ortolan Bunting
Emberiza hortulana
Rare autumn migrant. A juvenile at Lothbeg Point on 8 Sep 99 is the only recent record.

Yellowhammer
Emberiza citrinella
Breeding resident in the south and west. Following a marked decline and contraction of range towards the end of the last century, this species is making a welcome recovery, probably as a result of climate change. Winter flocks in the south-east can number more than fifty birds.

Pine Bunting
Emberiza leucocephalus
Vagrant. There have been no sightings since the winter records in 1976 and 1995.

Reed Bunting
Emberiza schoeniclus
Common breeding resident, wintering mainly on coasts. Flocks occur in the south-east, max. 30 in the Clynelish valley on 30 Dec 03. *See photo overleaf.*

Snow buntings are common on the coast in winter.

Yellow-breasted Bunting
Emberiza aureola
Rare autumn migrant for which there are two recent sightings. Just after dawn on 28 Sep 98, one arrived from the north at Durness, calling as it dropped down from a considerable height, and was later relocated near the village by birders looking for the missing Daurian starling. A juvenile was feeding with meadow pipits in weeds behind Brora's south-west beach on 18 Oct 02 and left with them to SW.

Little Bunting
Emberiza pusilla
Vagrant. On 9 Feb 03, a day of considerable mid-winter movement, one flew NE through the Clynelish valley, Brora, calling repeatedly. This regular autumn migrant on the Northern Isles occasionally overwinters in Britain, but is easily overlooked.

Rustic Bunting
Emberiza rustica
Vagrant, most likely in autumn. There have been no sightings since the singing bird at Baligill in mid-May 1991.

Reed buntings are resident but often retreat to the coast in winter.

Species Index

The page number of the annotated list account is given first, in bold type. Additional references in the text follow, in plain type. Photographs are in brackets; paintings in square brackets.

Species are listed under family or group names, with the individual descriptive names following in alphabetical order.

Auk, Little **155** 98, 102, 105
Auklet
  Cassin's **156**
  Crested **156**
Avocet **142**
Bee-eater **97**
Bittern **133**
Bittern, Little **133**
Blackbird **177**
Blackcap **172** 39, 101 (173)
Bluethroat **178**
Brambling **183** 95 (184)
Bullfinch **184** 38
Bunting
  Black-headed **188**
  Corn **188**
  Lapland **188**
  Little **190**
  Ortolan **189**
  Pine **189**
  Red-headed **188**
  Reed **189** (190)
  Rustic **190**
  Snow **188** 24, 58, 97, 104 (189)
  Yellow-breasted **190** 102, 106

Buzzard
  Common **139** (139)
  Honey **136** 96, 100, 117
  Rough-legged **140**
Capercaillie **128** 41
Chaffinch **183** 38, 43
Chiffchaff **172** 101
Chiffchaff, Siberian **172** 100
Coot **142**
Cormorant **133** 12
Corncrake **142** VII, 105
Crane
  Common **142**
  Sandhill **142** 102
Crake
  Baillon's **141**
  Spotted **141**
Crossbill
  Common **187** 41, 97, 103
  Parrot **187**
  Scottish **187** 41, 103
  Two-barred **187** 96
Crow
  Carrion **169**
  Hooded **169**

Cuckoo 164
Curlew 146 14, 20, 26, 48, 69 (21) [108]
Curlew, Stone (see Stone Curlew)
Dipper 176 30, 101
Diver
  Black-throated 129 VII, 50, 56, 71, 96, 101, 103 (129)
  Great Northern 130 95, 104, 106, 117
  Red-throated 129 V, 50, 96, 100 (51)
  White-billed 130 98, 102, 104
Dotterel 144 V, 57 (59)
Dove
  Collared 163 (164)
  Rock 163
  Stock 163
  Turtle 164
Duck
  Black 123 101
  Harlequin 126 99, 102
  Long-tailed 126 23, 100, 105, 106 (23, 93, 127) [116]
  Mandarin 122
  Ring-necked 125 105
  Tufted 125 100, 105
Dunlin 149 18, 48, 96 (20, 51)
Dunnock 181
Eagle
  Golden 140 V, 54, 62, 95, 103 (54) [114]
  White-tailed 137 95, 103, 117 (138)
Egret
  Great 133
  Little 133
  Snowy 133
Eider
  Common 125 22, 100, 101 (23, 126) [112]
  King 126 23, 102
Falcon
  Gyr 141
  Peregrine 141 62, 101 [115]
  Red-footed 141
Fieldfare 177 95 (178)
Finch, Trumpeter 185
Firecrest 169
Flycatcher
  Pied 179
  Red-breasted 179
  Spotted 178 103 (179)
Fulmar 130 12, 98, 103, 104, 106 [111]
Gadwall 122
Gannet 133 70, 98, 104 (132)
Garganey 124 104, 117 (124)
Godwit
  Bar-tailed 146 18
  Black-tailed 146 103 (146)
Goldcrest 169
Goldeneye 128 100
Goldfinch 188
Goosander 128 100
Goose
  Bar-headed 120
  Barnacle 120 10, 104, 105 (11)
  Brent 121 104
  Canada 120 100
  Greylag 120 21, 33, 45, 50, 69, 100, 101 (22) [112]
  Greenland White-fronted 120 104, 105

Lesser Canada 120
Pink-footed 119 VIII, 45, 69, 97, 103 [112]
Red-breasted 121
Ross's 120
Snow 120
Sushkin's 119
Taiga Bean 119 105
Tundra Bean 119 (118)
White-fronted 119 100
Goshawk 139 96, 103
Grebe
 Black-necked 136
 Great Crested 135 100
 Little 135
 Red-necked 135 101
 Slavonian 135 23, 100, 105 (136)
Greenfinch 185 45
Greenshank 151 VII, 29, 48, 56, 96, 101, 104 (151) [109]
Grosbeak, Pine 184
Grouse
 Red 128 53, 54
 Black 128 43, 103 (43) [113]
Guillemot
 Black 154 12, 14, 106 (14)
 Brünnich's 156 98, 102
Gull
 American Herring 161
 Black-headed 159
 Bonaparte's 159
 Common 160
 Glaucous 163 102
 Great Black-backed 163 9
 Herring 161

Iceland 162 102
Ivory 158 98, 102
Kumlien's 162
Laughing 160
Lesser Black-backed 161
Little 159 101, 102
Mediterranean 160 102
Mew 160
Ring-billed 161
Ross's 160 98, 102, 157
Sabine's 159 99, 102, 104, 105 (158)
Taimyr 161
Yellow-legged 161
Harrier
 Hen 138 95, 100, 101, 103
 Marsh 138 100
 Montagu's 139 100
 Pallid 138
Hawfinch 184
Heron
 Grey 134 14, 28 (134)
 Purple 134 104
Hobby 141 96, 104
Ibis, Glossy 135
Jackdaw 168
Jay 168 96, 101, 117
Kestrel 141
Kingfisher 166
Kite
 Black 137 97
 Red 137 46, 71, 95, 117 (137)
Kittiwake 159 12, 70, 98, 104, 106 (13) [111]
Knot 147 18, 101 (19) [108]
Lapwing 145 26, 29, 48, 69, 97 (145)

Lark
  Shore   170 58
  Short-toed   171
Linnet   185 45
Magpie   167
Mallard   123 32
Martin
  House   171
  Sand   171
Merganser, Red-breasted   128 32, 48, 101, 105
Merlin   141 54, 101
Moorhen   142
Nightingale, Thrush   96
Nightjar   165
Nutcracker   168
Nuthatch   175
Oriole, Golden   167
Osprey   140 71, 95, 101 (140)
Ouzel, Ring   176 54, 104
Owl
  Barn   164 71
  Long-eared   165 103
  Short-eared   165 43, 100, 103
  Snowy   164
  Tawny   165
Oystercatcher   143 14, 20, 26, 29, 48 (20, 143)
Partridge
  Grey   129
  Red-legged   128
Petrel
  Fea's   130 98, 99
  Giant sp.   130
  Leach's   133 102, 104, 105
  Storm   132 10, 102, 104
  Wilson's   132 99

Phalarope
  Grey   150 102, 105
  Red-necked   149 105
Pheasant   129
Pintail   123 100 (123)
Pipit
  Meadow   182 54
  Olive-backed   182
  Red-throated   183
  Richard's   182 100, 104
  Rock   183
  'Scandinavian' Rock   183
  Tawny   182
  Tree   182 100
  Water   183
Plover
  Golden   143 48, 54, 65, 66, 96 (51) [109]
  Greater Sand   145 100
  Grey   143 71, 100
  Kentish   145 100, 102
  Little Ringed   144
  Ringed   144 20, 48 (20, 144) [107, 113]
  Semipalmated   144
Pochard   125
Pochard, Red-crested   125
Ptarmigan   128 58 (59)
Puffin   154 V, 12, 13, 106 (14) [110]
Quail   128
Rail, Water   141 103, 117
Raven   169 62
Razorbill   154 12, 106 (155)
Redpoll
  Arctic   187
  Common (Mealy)   186 95, 117

Greenland   186
  Lesser   186   38 (186)
Redshank   152   26, 29, 48 (152)
Redshank, Spotted   151
Redstart   179   39, 100
Redstart, Black   179
Redwing   178   39, 95, 101 [114]
Robin   178
Roller   166
Rook   168
Rosefinch, Common   184   96
Ruff   147   95
Sanderling   148   71 (148)
Sandpiper
  Baird's   149   100
  Broad-billed   147   102
  Buff-breasted   149
  Common   150   56, 101
  Curlew   147
  Green   151
  Pectoral   149   105
  Purple   148   101
  Stilt   147
  White-rumped   149
  Wood   152   VII, 96 (92)
Sandgrouse, Pallas's   163
Scaup   125
Scaup, Lesser   125
Scoter
  Black   127
  Common   126   VII, 23, 50, 96, 100, 103
  Surf   127   23
  Velvet   127   100
Shag   133   12, 106
Shearwater
  Balearic   132   99, 104

  Barolo's   132
  Cape Verde   131
  Cory's   130   104
  Great   131   99, 104
  Manx   131   99, 104
  Sooty   131   99, 104
Shelduck   121   26, 101 (26, 121)
Shelduck, Ruddy   122
Shoveler   124   117
Shrike
  Great Grey   167
  Lesser Grey   167
  Red-backed   167   96
  Woodchat   167   106
Siskin   188   38
Skua
  Arctic   154   9, 106 (10)
  Great   154   9, 106 (10)
  Long-tailed   154   104
  Pomarine   153   104
  South Polar   154   102
Skylark   170   54, 97
Smew   128   103, 104
Snipe
  Common   153   29 (153)
  Great   153
  Jack   152
Sparrowhawk   139
Sparrow
  House   181
  Tree   181 (181)
Spoonbill   135
Starling
  Common   176
  Daurian   176   106
  Rose-coloured   176
Stilt, Black-winged   142

Stint
 Little 149
 Temminck's 147 VII, 96 (92)
Stonechat 180 45, 101
Stone Curlew 142 98
Stork
 Black 135 101
 White 135 97
Swallow 171
Swallow, Red-rumped 171
Swan
 Bewick's 119 105
 Black 119
 Mute 119
 Whooper 119 21, 28, 95, 103, 105, 117 (93)
Swift
 Common 165
 Pallid 165
Teal
 Blue-winged 124
 Common 123 23, 32, 48, 56
 Green-winged 123
Tern
 Arctic 158 106
 Black 157 100
 Bridled 157
 Common 158
 Little 157
 Roseate 158
 Sandwich 158 (157)
Thrush
 Black-throated 177
 Mistle 178
 Song 177 45
 White's 176
Tit

 Blue 169 38
 Coal 169 38, 43
 Crested 169 41
 Great 169 38
 Long-tailed 171 38
Treecreeper 175 38
Turnstone 147 (147)
Twite 185 103
Wagtail
 Grey 182 30
 Pied/White 182
 Yellow (all races) 182
Warbler
 Barred 173
 Blyth's Reed 96
 Booted 174 102
 Garden 173 96, 103
 Grasshopper 174 96
 Icterine 175 96
 Melodious 175
 Pallas's 171 106
 Reed 175
 Sedge 175 103
 Subalpine 174
 Willow 172 38 (172)
 Wood 171 38, 100
 Yellow-browed 171
Waxwing 175 97 (174)
Wheatear
 Desert 180 106
 Northern 180 106 (180)
Whimbrel 146 95
Whimbrel, Hudsonian 146
Whinchat 179 103
Whitethroat
 Common 174 39
 Lesser 173 96, 103, 117

Wigeon   **122**   17, 28, 32, 48, 101 (18, 32, 122)
Wigeon, American   **122**   105
Woodcock   **152**   38
Woodpecker
  Great Spotted   **167**   38 (39)
  Green   **167**   101
Woodpigeon   **163**
Wren   **176**   (177)
Wryneck   **166**   96 (166)
Yellowhammer   **189**   24 (24)
Yellowlegs, Greater   **151**   101 (150)

References

Baxter, E V & Rintoul, L J. 1953. *The Birds of Scotland*. Oliver & Boyd

Stroud, D A., Reed, T M., Pienkowski, M W., & Lindsay, R A. 1987. *Birds, Bogs and Forestry*. Nature Conservancy Council

Forrester, R & Andrews, I. 2007. *Birds of Scotland*. Scottish Ornithologist's Club

Thom, V M. 1986. *Birds in Scotland*. Poyser

Vaurie, C. 1959. *Birds of the Palearctic Fauna Vol.I*. Witherby

Vittery, A. 1997. *The Birds of Sutherland*. Colin Baxter

Vittery, A. 2001. *Little Gulls feeding in association with auks*, Scottish Birds 22:107–108

Vittery, A. 2005. *Food-induced erythrism in House Sparrows*. British Birds 98:433–435

Whitfield, D P. 1997. In *Conserving Peatlands*. Ed. Parkyn, Stoneman and Ingram. CAB International

Acknowledgements

The authors are very grateful for the assistance of Dean MacAskill who has spent many hours in the field and provided numerous records of birds. Without this contribution the book would be incomplete.

Tony Mainwood supplied valuable data on Twite and assisted in many other ways. Judith Anderton read the draft manuscript and provided useful comments on the text. Many others contributed indirectly by submitting ornithological records over the years. To all, we are most grateful.

The Authors

Fraser Symonds developed a passion for birds at an early age and was painting and photographing them by the age of ten. He qualified for a bird ringing licence in the 1970's, whilst still a teenager. After obtaining a Zoology degree from Durham University he went on to a career in wildlife conservation, initially with the Nature Conservancy Council undertaking research on wading birds and latterly with Scottish Natural Heritage. Since 1989 he has lived near Dornoch in east Sutherland and took early retirement in 2014 in order to spend more time on painting and photography.

Alan Vittery has been a serious bird-watcher for 65 years and moved to Brora in Sutherland in 1990 after taking early retirement from the Nature Conservancy Council. Now a full-time birder, he was able to realise the potential of this under-watched County, finding many scarce migrants more associated with the Northern Isles. Intensive seawatching over a period of twenty years, from both the east and north coasts, put both Brora and Strathy Point firmly on the ornithological map and has demonstrated that the Moray Firth ranks as one of the most important seabird 'capture zones' in the British Isles. He published *The Birds of Sutherland* in 1997.